Books by William Stockton:

ALTERED DESTINIES

FINAL APPROACH: THE CRASH OF EASTERN 212

ALTERED DESTINIES

ALTERED DESTINIES

William Stockton

1979
DOUBLEDAY & COMPANY, INC.
GARDEN CITY, NEW YORK

Library of Congress Cataloging in Publication Data

Stockton, William.
  Altered destinies.

  1. Medical genetics. I. Title.
RB155.S89   616'.042
ISBN: 0-385-14011-8
Library of Congress Catalog Card Number 78-73196

COPYRIGHT © 1979 BY WILLIAM STOCKTON
ALL RIGHTS RESERVED
PRINTED IN THE UNITED STATES OF AMERICA
FIRST EDITION

For

ANN CECILIA SNYDER

INTRODUCTION

Lightning only strikes once!

That's what a patient of mine was told—not once, but twice! This woman's story is as tragic as it would have been avoidable had she or her physician been well informed. She appeared in my office at the Albany Medical Center with a troubling question: did her three-year-old son have muscular dystrophy? The question had arisen because she had a twelve-year-old son who was bedridden with the disease and an eight-year-old son confined to a wheelchair with MD. But on each occasion, when the crippling and ultimately fatal disease was diagnosed she was told not to worry. Lightning only strikes once. She had been told the chance of a third MD child was unheard of. But in truth, there was a 50 percent chance that each son she bore would have the disease, and a 50 percent chance that each daughter would be a carrier of the disease but not affected by it. If only she had known . . .

When the laboratory results came back on the three-year-old, I had to tell the woman the awful truth: her third son also was destined to live the life of his two older brothers. By age six he would be in a wheelchair. At age twelve he would be bedridden, and by twenty he would be dead. Every day she would have to face her three crippled boys alone. Their father had left the family after the second son's diagnosis had been confirmed.

There are more horrors associated with this woman's story. Her brother had also had MD. All during her childhood, she had witnessed the progressive development of the disease in him. She had to help with his care, carrying him up from the sofa to his bed every night until he finally died at eighteen. Her own development had been woefully neglected as a result. She always came second. She looked back on her childhood with nothing but resentment. Now, she told me, her first thought every morning was suicide!

This is a frightening story from many points of view, both medically and socially. But to me, as a geneticist and a physician, it is particularly sobering because it illustrates an important responsibility of our medical specialty: teaching our colleagues and our patients.

Teaching, in fact, is an essential part of a physician's life. It is indoctrinated in us from the time we enter medical school, it is stressed in the Hippocratic oath that we take on graduation, and it should be part of our practice of medicine. It is a debt we owe our teachers and the patients upon whom we practiced.

The obligation to teach is a tradition in the medical profession. Indeed, the word *doctor* is derived from the Latin verb *doceo*, meaning "teach." As Lewis Thomas points out in *The Lives of a Cell:*

> The antique word for physician is leech. It is also the word for the worm *sanguisugus*, used for leeching. Leech the doctor means the doctor who uses leech the worm; leech the worm is a symbol for the doctor. Leech the doctor comes from the Indo-European *leg*, which meant to collect, and by derivation, to speak. *Leg* became Germanic *lekjaz*, meaning one who speaks magic words, an enchanter, and also *laece* in Old English, meaning physician. *Leg* in its senses of gathering, choosing, and speaking gave rise to the Latin *legere*, and thus words like lecture and legible. In Greek, it became *legein*, meaning to gather and to speak; legal and legislator and other words derived. *Leg* was further transformed in Greek to *logos*, signifying reason.

Because of this tradition, the standard of teaching in medical schools is high and considerable kudos is attached to the great teachers in the profession. Yet physicians are peculiarly inept at translating their knowledge and experience into a form that the public can understand. And this is even more startling when the avowed aim of many branches of medicine now is greater emphasis on teaching the public, because of the broadening of the basis of medicine to include the concept of prevention.

It is our life-style that causes us the most ill, according to the heart specialists, nutritionists, and liver experts who tell us how to prevent high blood pressure and heart attacks, lung cancer and alcoholism. We sit, we smoke, we drink, and we eat too much. If we climbed a mountain every day and ate yogurt at night, we would live longer and we would be healthier during the time we lived.

Medical genetics has been in the forefront of this movement of prevention. As a matter of fact, the principal aim of genetics is prevention: prevention through reproductive options for couples who are at high risk for having affected children, prevention of Rh and rubella babies by immunization, and prevention of mental retardation by detecting affected babies through newborn screening programs before they show manifestations of the disease.

But such prevention isn't possible unless the public is educated. And by our own admission we have not done a good job of educating the public. (Many of us also think we have done just as bad a job of teaching our colleagues.)

Why are we so bad at teaching the public? After all, we spend a lot of time telling our patients what is wrong with their child, what is his or her future, why they have a defective child, and why they might be at risk for having more affected children.

I think there are four reasons. One is that it is difficult to learn a subject in one language—the technical jargon—and then translate it into another—everyday language. It takes a lot of effort. Second, I think we must admit that one can more easily hide the ghastly truth behind technicalities, whereas it would be frightening for the physician, let alone the patient, if the truth were revealed nakedly in simple words. This is poignantly brought out in the story of Marsha and Gary and their Tay-Sachs baby later in this book.

Third, I think there is an element of self-consciousness. The student perceives that medical jargon lends an air of knowledge, and the established physician perceives that his language bestows on him a mantle of wisdom. If he simplifies his subject, perhaps his colleagues will accuse him of superficiality rather than revering him for his profundity.

Fourth, I think professional acclaim is an important reason. With a limited amount of time, one accrues more acclaim from one's colleagues by writing and talking to them in their language rather than by writing and talking to the public in theirs.

I mentioned owing our patients something. Remember, the medical profession is the only human activity where every attempt to relieve suffering is accompanied by the risk that we may also cause suffering. I am, of course, not talking about injections and unpleasant diagnostic procedures, but about something we have never dared explain to the public: namely, that good medical judgment is learned from making bad medical judgments. Yes, a little judgment can be learned from one's teachers and some from the experience of others, but most good medical judgment and clinical skill is learned the hard way— from practice. And it is not only students and interns who practice on people. We all do. We all know of patients who would have been alive today or who would not have had retarded children if we could have seen them at a time in our career when we were more experienced. This is illustrated in these pages by the story of Sandy in San Francisco, which has a happy ending because a senior physician stepped in before the learning by practice had gone too far.

Even after we have mastered our particular field, or reached the height of our own experience, we may still be the cause of our patient's suffering—not from a personal ignorance about the patient's disease, but from general ignorance. The patient suffers because the entire profession doesn't understand the disease, as in many birth defects and mental retardation.

It is time the public understands the uncertainties of medical practice. Once we recognize that all our efforts to relieve suffering might on occasion cause suffering, we are in a position to learn from our mistakes and appreciate the debt we owe our patients for our education. It is a debt that we must repay—it is like tithing.

And we can repay the debt in many ways. We can be assiduous about keeping our knowledge and skills up to date, be available to patients at all hours, teach, or carry on research. The physician who attempts to combine investigation with a clinical career travels perhaps the toughest road. His counterpart in basic science thinks he is a dilettante researcher, his clinical colleagues think he is unsafe, and his mother-in-law says, "He's thirty-five years old and still working with animals. When will he be a real doctor?"

We have many opportunities to tithe. Compassion is one. But there are also other ways in which we can repay our patients for our education and feel we are earning the right to practice medicine. We can swallow our pride and discard our self-consciousness. We can practice what we preach. We can teach the public—lecturing lay groups, writing for their magazines, appearing on television and radio—and, finally, we can provide our specialized knowledge to professional writers. In some ways, this combines the best of two worlds: accuracy and literacy. It is in this spirit that ALTERED DESTINIES has been produced.

As a result of our poor track record in communicating with the public, some important misconceptions have arisen. For example, many people think that the human genetic constitution has been weakened by the advent of modern civilization. It seems only reasonable, people say, that if the environment softens, we too must soften. There is nothing new about this. In the 1890s it was observed that not only was the human jaw diminishing, but so also was the human intellect. And yet we have continued to witness the unsurpassed excellence of scientific and other achievements in the 1970s.

One element in this attitude can be traced to the romantic revival in the nineteenth century. The myth is that before the industrial revolution human beings lived in a state of happy innocence—vigorous, healthy, robust, and having a sort of inner tranquillity and radiance that cynics say comes nowadays only to those possessing a substantial private income. But this, of course, is nonsense. It is now generally agreed that the foothold of humankind on this earth was until a few hundred years ago not an entirely secure one.

Is there indeed any reason to suppose that advances in medicine are undermining the fitness of the human race? This argument is based on the belief that there is a genetic element in all human disease. This is known to be true of some diseases and is not known to be false of any, so here there can be no disagreement. But the argument then runs as follows: because of the discovery of, for example, insulin and antibiotics, we are preserving for life and reproduction people who previously might have died. We are, therefore, preserving the genetically ill-favored, the hereditary weaklings. They can marry and undermine the constitution of future generations, and as a result of this, human-kind is going downhill.

If by "going downhill" we mean declining biological fitness, with the implication that humankind will probably die out, this argument is full of contradictions. If medical treatment confers fitness upon the unfit, there can be no fear of extinction; and if it fails to do so, the fear of extinction does not arise.

The "going downhill" argument seems to contemplate the predicament of the modern human race in primitive surroundings without insulin and penicillin and other allegedly debilitating devices. It is true that we preserve for life people who previously might have died. It is also true that if some disaster were to destroy the great pharmaceutical industries to which diabetics and patients with many other diseases literally owe their lives, then a great many of them might die. But what could be deduced from this except the shameful inference that people who might conceivably die tomorrow might just as well be dead today?

For those who like more concrete examples, phenylketonuria will illustrate the point. At the present, about one in 10,000 newborn children has this disease. If it is untreated, its grimmest manifestation is mental deficiency. But now it is treatable. If these treated people have children, it will take forty generations, or more than one thousand years, to double the incidence of PKU. The projected incidence in the year 3000 will be one in 5,000. Big deal! How many people each year receive serious brain damage in automobile accidents and are neurologically similar to or worse than a PKU victim?

Some people believe that the human race might be improved through the application of simple genetic laws. But the human race cannot be improved by applying the methods of the professional stockbreeder. Indeed, the whole discussion of eugenics has been bedeviled by the false analogy between artificial and natural selection. Artificial selection is intensive special-purpose selection aimed at producing a particular excellence, whether in milk, speed, or fancy images in dogs. It produces a number of specialized pure breeds, each less varied than the parents. But we owe our evolutionary success to our variability.

I think it is important to understand that we are all different genetically—each of us is unique (with perhaps the exception of identical twins). Not only are we all different now, but everyone who ever lived was and will always remain unique. Some of these differences are for the good, some may cause birth defects, and some may predispose us to acquiring diseases in later life. But most differences appear to have neither particularly good nor particularly bad effects. The important point is that each one of us is, in a sense, potentially the carrier of a gene that in a given circumstance may be responsible for the survival of our species. Our diversity as a species is in effect a reservoir of possibilities to respond to the numberless environmental changes that are and will be taking place, an insurance policy against extinction.

As a result of lack of public awareness and knowledge, genetics took a strange and shameful course in the 1930s. In 1865 Gregor Mendel presented a paper, "Experiments in Plant-Hybridization," to the Society of Natural Science in Brünn, Moravia, then a quiet corner of the Austrian Empire. In his paper he proposed, as all the world now knows, the idea of a simple, highly regular algebra of heredity in which discrete units —he called them elements; the word *gene* came later—combine and recombine down the generations. It is the cornerstone of genetics. That same year, Sir Francis Galton wrote a paper entitled "Possible Improvement of the Human Breed." The thoughts expressed in this paper were the basis for what came to be called the eugenics movement, which during the 1920s and 1930s became so dominated by a missionary zeal to improve the human race that it disregarded scientific methods, using precarious claims to support sweeping recommendations for the betterment of humanity. In fact, many of the studies were little more than accessories made to advance the author's preconceived eugenic notions. The movement tended here to become science, there a social movement, and in Germany it became an application of prejudice through the so-called eugenics laws of 1933. Eugenics in Germany (and in other countries) was achieved mostly by sterilization of "undesirables" and was enforced by an elaborate legal mechanism, which was a matter of common public knowledge and had at least the tacit approval of most of the legal and medical profession. Medical geneticists played important roles in the implementation of these laws by investigating family histories, giving expert evidence in the courts, and participating in the eugenics courts proceedings.

The first step in the legal process was taken by the district health officer. If a mentally retarded or otherwise defective person was brought to his attention, it was his responsibility to investigate the case and, if appropriate, in his opinion, submit a proposal for sterilization to the local eugenics court. This court, consisting of a lawyer and a medical geneticist, reviewed the proposal and either rejected it or recommended sterilization. If the person in question disagreed with the local court's decision, he or she was entitled to an appeal to a higher eugenics court whose decision was final.

Coupled to this legislation was the Law for the Protection of the Genetic Well Being of the German People, which banned marriages when one of the parties was regarded as unsuitable because of mental disorder, inherited disease, or racial origin.

The full horror of these proceedings, let alone the tragic impact on individuals and their families, had a far-reaching effect on the development of genetics both as a scientific discipline and as a medical specialty. Geneticists have a deeply felt concern about patient confidentiality as a result. That's why the geneticists in Seattle were so concerned about privacy in the case of Christine, the adopted child at risk for myotonic dystrophy, whom you will meet later in these pages.

American medicine paid little attention to genetics during the first three decades of this century. In fact, few American medical schools provided any formal instruction in genetics. This was in part owing to the adverse impact of the eugenics movement and in part owing to the prevailing view among physicians that inherited diseases were academic playthings of no practical importance. Hereditary conditions were so rare that most physicians saw not a single one during an entire career. But gradually the practice of medicine became more receptive to development in genetics. This was partly because infectious diseases, which until then had been the most important cause of death, now produced a less significant proportion of infant deaths. Children were better nourished and protected by one means or another against infection. The death rate was lower, and the expectation of life at birth was increasing. At the turn of the century, tuberculosis, pneumonia, and gastrointestinal infection accounted for 75 percent of all deaths. Now they account for only 10 percent. In contrast, approximately 4 percent of infant deaths were then attributed to birth defects; now about 20 percent are. To put it another way, in 1900 one infant death in approximately 25 was due to congenital malformations; now it is about one in 5. Medical school curriculum committees, therefore, could no longer argue that instruction in genetics would overload the curriculum with material the future practicing physician would not use. With a knowledge of heredity, the physician could counsel prospective parents with genetic conditions and thereby prevent the unwitting passage of a disease to subsequent generations.

Genetic counseling was, in fact, not always part of medical practice. As late as 1972, only three fourths of the medical schools in America had courses in genetics. There is a striking difference between those physicians who graduated more than ten years ago and more recent graduates in terms of their knowledge of genetics. Physiology, biochemistry, and molecular biology were developed in medical schools, and students were exposed to the newest ideas as they unfolded, but genetics was developed by zoologists and botanists, some of whom—unlike their clinical colleagues—saw its relevance to human beings. Genetic counseling was begun by nonmedical geneticists or by physicians trained in genetics but working in departments or clinics not directly engaged in patient care. Thus genetic counseling developed apart from clinical medicine.

One traditional principle of genetic counseling is the neutrality of the counselor in decisions about reproduction. Unlike the heart specialist, who usually *tells* the patient what the treatment will be and orders him or her into the cardiac care unit, the geneticist offers the patient a choice. This is unusual in medical practice. It is a difficult attitude for many physicians to adopt and may even be confusing to some patients, who expect to be guided by their physician. This attitude of neutrality may have originated in counselors not engaged in patient care, who may have felt some reluctance, therefore, to enter into the lives of their counselees in the way practicing physicians do. In addition, there were the inhibiting examples of the abuses of eugenics in the previous decades. With prolonged experience and with more physicians providing genetic counseling, we may expect a change to the more traditional doctor-patient relationship.

Because we humans possess minds, we have become the agent responsible for our evolution. It will only be by the right use of our minds that we'll be able to exercise that responsibility. We will only succeed if we face it consciously and use all our mental resources—knowledge and reason; imagination and sensitivity; capacity for wonder and love, for comprehension and compassion, for spiritual aspirations and moral effort.

Our development occurs by two distinct methods:

—the genetic, based on the transmission and variation of genes, and
—the cultural, based on the transmission and variation of knowledge and ideas.

Biological heredity is handed down from parents to children at the moment of conception by means of the genes. Our cultural heritage is transmitted by people regardless of their relationship by means of teaching and learning. Cultural evolution is very much faster and in many ways more efficient.

The Lamarckian theory, until quite recently popular in Russia (and responsible, I might add, for the poor state of Russian agriculture), declares that the environment can somehow issue genetic instructions to living organisms, instructions which, duly assimilated, can be passed on from one generation to the next. There is, of course, a tremendous psychological inducement to believe in this theory of evolution, but any analysis of what has appeared to be a Lamarckian style of heredity has shown it to be non-Lamarckian. The environment does not imprint genetic instructions upon living things.

Lamarck illustrated his notion of the inheritance of acquired characteristics with the giraffe. Suppose that a primitive antelope-like creature that fed on leaves ran out of food within easy reach and had to stretch its neck as far as it could to get more food. By habitual stretching of its neck, tongue, and legs, it would gradually lengthen those appendages. It would then pass

on the developed characteristics to its offspring, which in turn
would stretch farther and pass on a still-longer neck to their descendants, and so on. Gradually, after generations of stretching,
the primitive antelope would evolve into a giraffe. If this were
the mechanism, it is difficult to explain the giraffe's blotched
coat, for surely no action on its part, deliberate or otherwise,
could have effected this change.

A natural experiment of this kind has been taking place for
more than a hundred generations in Jewish boys, and yet there is
no evidence that circumcision has led to any shriveling of the
foreskin.

Blacks do not have dark skin because they live in Africa. It
would be truer to say that they have thrived in Africa because
they are dark. The former would be an example of Lamarck's
"inheritance of acquired characteristics"; the latter, an example
of natural selection. Dark skin has obvious value in tropical and
subtropical climates, just as fair skin is useful in colder climates
to absorb as much ultraviolet radiation as possible from the com-
paratively feeble sunlight. Likewise, the protected eyes of the
Eskimo and Mongol have survival value where the glare from
snow or desert is intense, and the high-bridged nose and narrow
nasal passages of the European serve to warm the cold air of the
northern winter.

The environment brings out the genetic potential present in
the embryo, but it does not instruct the developing embryo in
the manufacture of its particular enzymes or proteins. Those instructions are already embodied in the genes. The environment
causes them to be carried out. What humans are is genetically
determined. What they do is culturally determined.

But human cultural activities do, of course, have a profound
influence upon the physical environment and in turn upon our
genetic makeup. Infectious disease, for example, can play a vigorous role in such interactions. The slash-and-burn hypothesis
provides a notable instance when human activities favored the
spread of malaria. With the introduction of Iron Age tools, the
jungle was ripped open and the conditions thereby created that
favored the endemicity of malaria. In these circumstances, people possessing intrinsic resistance to malaria would be favored.
Those who did not would die. Accordingly, genes that rendered

people who carried them more resistant spread in the population. Thus, humans, through their inventions, altered their genetic constitution. This argument has been extended to include other infections such as plague, typhus, smallpox, cholera, and tuberculosis.

Since the domestication of human beings five to ten thousand years ago, infectious disease has been by far the most important cause of death. Specific simple genetic alterations are present in some people that confer some degree of natural or inborn immunity to infections, but there is not the slightest reason to believe that such alterations are of any genetic virtue or confer any advantage whatsoever when those hazards are not present. In general, we cannot say that improvement of the environment improves humankind genetically.

In brief, then, biological evolution is given direction by the blind and automatic agency of natural selection operating through biological mechanisms; but human evolution is given direction by cultural guidance operating with the aid of mental awareness, notably the mechanism of reason and imagination.

The Catholic priest Teilhard de Chardin saw that the evolution of matter, the evolution of life, and the evolution of the human race are integral parts of a single process, a single and coherent history of the whole universe. He said in *The Phenomenon of Man:*

> Man is not the center of the universe as was naïvely believed in the past. But something more beautiful—man is the ascending arrow of the great biological synthesis. Man is the last born, the keenest, the most complex, the most subtle of the successive layers of life. This is nothing less than a fundamental vision.

Finally, the most important point of Darwin's teaching was strangely enough overlooked. This is that the human race has not only evolved, but is evolving. We humans and we alone know that the world evolves and that we evolve with it. By changing what we know about the world, we change the world that we know. By changing the world in which we live, we change ourselves.

During the last twenty-five years, progress in genetics has been rapid and breathtakingly exciting: the genetic code has been deciphered, and the details of protein synthesis and regulation are being worked out, partly by brilliantly formulated hypotheses and partly by some extraordinarily sophisticated experiments. James D. Watson, in his memoir *The Double Helix*, was hardly shy about the significance of the discovery of the structure of DNA. He called it "perhaps the most famous event in biology since Darwin's book." (Darwin published *The Origin of Species* in 1859.) Judging by the reports of these developments in the popular press, they are having the same impact upon our generation as Darwin's hypothesis had on his.

The recent flowering of molecular genetics is, indeed, an astonishing phenomenon. Before 1954 only two Nobel prizes for physiology or medicine had been given to geneticists. Since 1954, fourteen molecular geneticists have received Nobel prizes.

Geneticists, chemists, and physicists now discuss biologic problems in the common language of molecular structure. Molecular genetics has emerged as the leading discipline in biology, in much the same way that nuclear physics emerged as the leading branch in physics in the 1930s. The tone of physics was then—as it is in biology now—creatively confident, argumentative, lavish, and full of hope. The tone of science in the 1930s was that of Ernest Rutherford, Niels Bohr, and Albert Einstein. Now it is that of the molecular biologists, François Jacob and Jacques Monod, James D. Watson and Francis Crick and Linus Pauling. To be sure, the discoveries have not produced the great practical payoff that has so long been anticipated for them. No baby has been cured of a congenital deficiency by insertion of a missing gene into its cells. There is no vaccine against human leukemia. There's not even a cure for hay fever. Though some of the rewards are at least imminent, most scientists have learned that they must speak guardedly and emphasize to the public the gaps in our knowledge that need to be filled in by more research.

The most difficult problems are yet to come; they may prove to be the most interesting. We are certain to witness some very exciting developments in the near future. We can only hope that we will be better prepared to meet the responsibilities that will result than we were to meet those that arose out of the development of atomic theory.

Whether this happens will depend upon the public's eagerness to learn, which in turn depends upon the geneticists' willingness to teach. We geneticists have an obligation to you, our public. We all have an obligation to society.

The patients and their families you will meet in this book—whose destinies have been altered by their genes—are both the benefactors and the beneficiaries of medical genetics. They are a testament both to how far and how quickly we have come and to the further promise that medical genetics holds for us all.

Stanley Hoostowski, the stationmaster for the New Haven Railroad in New Canaan, Connecticut, said: "Progress is getting ahead of everybody." Jacques Monod, a theoretician in molecular genetics at the Pasteur Institute in Paris, answered a question in mild surprise: "The secret of life? But in principle we already know the secret of life."

Both men are right. We probably do know the secret of life—in principle. And progress will get ahead of everybody only if we let it, through our own lack of interest and neglect.

Ian H. Porter, M.D.

AUTHOR'S NOTE

The patient's privacy is jealously guarded in medicine. But most physicians are willing to introduce a writer to an interesting patient if there are assurances that the patient's identity will be protected. In the course of gathering material for this book, I spent much time with a variety of genetics patients throughout the country. With one exception, every family asked to have its identity disguised. So I have changed names, sometimes been vague about a job or a neighborhood, and even created a new town in west Texas. But every person is real. Every problem, every frustration, every candid remark, every feeling is genuine. There is no "faction" here, no melding of several patients into one.

The one case where the names are real is that of The Singing Halls. The Hall family and I were driving down a highway in South Carolina one Saturday when I broached the subject of disguised identities. Jerry Hall twisted around in the driver's seat and looked me in the eye. "If you don't use our real names, I'll stop this old bus and put you off in one of them swamps, Yankee boy," he said. He was joking about the swamp, but not about using his name.

WS
December 1, 1978

ALTERED
DESTINIES

AN AFTERNOON
IN BAR HARBOR

When the door at the front of the auditorium opened and the little retarded boy with the shock of red hair stepped through, everyone in the audience involuntarily leaned forward at the sight of his misshapen form. Their movement caused the padded theater-style seats to squeak loudly. The child, who was about ten, saw the sea of faces, heard the frightening squeaking, and sensed something threatening in the shifting of so many bodies toward him.

He whirled and would have fled back through the door, but his mother was in the way. He crashed into her body. The doctor gently took him by the shoulders and turned him about, steering him to a folding chair beside the podium. The boy's name was Raymond. He was the first genetics clinic patient of the afternoon.

My afternoon in Bar Harbor had its beginnings in 1929. That was not a year when the president of a major university could divorce his wife and hope to retain his job or place in the community. So when Clarence Cook Little decided to end his marriage, he sought a refuge far away from the University of Michigan and the town of Ann Arbor. He chose Bar Harbor, a posh resort community of enormous summer mansions on a small island off the coast of Maine. Little was an eminent biologist, an expert in genetics. He founded a small laboratory to specialize in genetics research, now known as the Jackson Laboratory.

The Jackson Laboratory endured the Great Depression, World

War II, and a fire that burned it to the ground in 1947. It
prospered, specializing in the use of highly inbred strains of mice
to attack fundamental biological problems that range from cancer to heart disease. Jackson is now considered the largest center
for mammalian genetics research in the world.

Each summer the National Foundation–March of Dimes and
Johns Hopkins University sponsor a two-week genetics "short
course" for doctors who want to learn about medical genetics.
The March of Dimes supplies the money and Johns Hopkins provides the faculty. The scientists meet in morning and evening
sessions, their afternoons free for sailing, hiking, tennis, and
other sports. They listen to one another present the latest research results in genetics and they discuss new methods to study
current problems. The emphasis of the short course alternates
each year. One year the sessions are devoted to mouse genetics;
the next year, human genetics.

For more than twenty years the guiding light behind the an-
nual sessions has been Victor McKusick, now the chairman of
the department of medicine at Johns Hopkins. If any medical
geneticist in the United States today were to be singled out as
the most important figure in the field, it probably would be
McKusick. He is a tall man with a receding hairline and a shock
of usually disheveled white hair. A small bemused smile plays
over his lips most of the time, and his seemingly bumbling nature makes him the archetypal absentminded professor. There is
genuine but fond disagreement among his students and colleagues about whether perhaps it is all a pose. No matter,
though. He is at once warm, ingenuous, charming, and very
much in command of every situation in which he finds himself.

McKusick is the author or co-author of more than one thousand scientific research papers, a prodigious number by any
measure. He has trained dozens, perhaps hundreds, of medical
geneticists. It is impossible to go into any medical genetics center in the country and not find a physician who either studied
under McKusick or studied under someone who in turn studied
under McKusick. But McKusick's place in the history of medical
genetics will be assured if for no other reason than his landmark
medical text: *Mendelian Inheritance in Man.* In this epic, McKusick catalogs all of the more than 2,300 known medical condi-

tions that can be passed genetically from one generation to another. It is a compendium of all the possible genetic horrors that can be visited upon a family, the worst nightmares of every pregnant woman who worries that her baby won't be normal.

When the short course is devoted to human genetics, a highlight of the session is the afternoon when McKusick presents patients he and his colleagues at Johns Hopkins have seen in their genetics clinics. Many of the doctors forgo their recreation to attend the session, perhaps as much to watch McKusick's presentation as to see the patients. It was in early August of 1978 that I attended the clinical session in Jackson Laboratory's auditorium, a room shaped like a wedge of pie with tiers of seats climbing toward the back, brown carpet, and comfortable upholstered theater seats, each with a small writing desk that folds away. McKusick was there, dressed in a tan corduroy coat and red plaid pants, playing his combined roles of impresario, master of ceremonies, and teacher.

The genetics patients were all children or adolescents, except for a father and his three daughters suffering from a condition called Charcot-Marie-Tooth disease, which involves muscle weakness and atrophy in the lower legs. All the children were accompanied by their mothers, although occasionally a father was present, too. In a few cases, other relatives sat in the audience. The patients and their mothers entered the room case-by-case through a door behind the lecturer's podium at the front of the room. Only one family was permitted in the room at a time. Before they entered, McKusick outlined the nature of their condition, sometimes showing slides on a rear projection screen.

I arrived early, reserving a seat in the auditorium with my notebook and tape recorder and then wandering out into the laboratory's lobby to watch the doctors arrive. I quickly realized that the patients and their families also were arriving. They were escorted into a side room where soft drinks were being served. I watched one obviously retarded and deformed little boy come into the lobby with his mother, a woman of middle age. He was the little boy who later would try to flee. Besides his red hair, he had a large, square head with a flat nose and large protruding lips. I could see the outline of a brace beneath his shirt. Its obvious purpose was to correct his severe scoliosis, or curvature of the spine. He walked with a bowlegged gait that was almost a stumble. He muttered to himself.

I sat down on a couch with a newspaper open in my lap. A few minutes later, a young woman, who I presumed worked for the laboratory, came out of the side room with the redheaded boy. She took him on a tour of the lobby, stopping to admire a display case containing several cages of live mice. Then they sat down on my couch.

I studied the newspaper intently, not wanting to stare, torn between the desire to watch the boy's every move and the desire to rise and flee back into the auditorium. The child was not as retarded as I first thought.

"What is this place?" he asked the young woman.

"It's a laboratory," she answered loudly. He apparently was hard of hearing also.

"Laboratory. Laboratory. Laboratory," he began chanting loudly, emphasizing each syllable of the word. "Laboratory. Laboratory. Laboratory." He repeated the word again and again for nearly a minute, until I was ready to leap up and hurry into the auditorium, regardless of how it might appear.

Just then the girl interrupted. "Raymond?" she said sweetly. I stole a glance at her. She had a look of serene tolerance on her face. "Do you know what they do here?"

The question diverted him. "What?"

"They do experiments on mice."

He looked at her blankly for a few seconds and then seized on the word "mice." He began chanting again. "Mice. Mice. Mice. Mice. Mice." He rocked his body to and fro with each word.

The lobby was filling. It was almost one-thirty, the scheduled starting time. I folded the newspaper, laid it carefully beside me, and rose and walked casually into the auditorium. Raymond's voice followed me, reaching through the growing noise of the crowd into the auditorium. I thought of Raymond's mother, probably enjoying a soft drink in the nearby room, thankful that someone whose patience hadn't been worn thin would entertain her son for a few minutes. Did Raymond's mother love him? I wondered how well she dealt with her burden. It was a question that had bothered me often as I had gotten to know other families with genetic problems.

I particularly wanted to attend the clinical session because I was seeking another perspective on genetic disease and the doctors and patients who struggle with it. For more than a year I had been traveling about the country spending time with families afflicted with genetic disease, talking to their doctors, searching for the essence of what it means to have an error in your own genetic code.

I had come to know the facts and figures well. The families I had seen brought them to life.

I knew that one in every 10 live-born infants has a genetic disease or handicap, which means that as many as 20 million Americans have or will develop a genetic problem. One in every 3 admissions to a pediatric hospital is for a genetic defect, and 40 percent of all infant mortality in the United States is due to genetic errors. The potential for alleviation of human misery is substantial if the results of modern genetics can be applied to the needs of every person with a genetic problem.

In fact, genetics is probably one of the fastest-growing, "hottest" areas in medicine today. This is because of fortuitous timing in basic advances in the technique and technology of medicine as well as fundamental research discoveries about the genetic code itself and how it governs the body.

The development of amniocentesis in the 1960s was a turning point in offering meaningful genetic services to patients. The procedure involves inserting a needle through the pregnant woman's abdomen to withdraw a sample of the fluid in which the growing baby floats in the amniotic sac. The procedure would have had little meaning unless laboratory procedures were also developed to analyze the fluid and spot birth defects in the making. The list of such analyses and of the diseases they can spot in time to permit a therapeutic abortion is growing.

Growing also is the list of discoveries about the nature of various hereditary diseases and the development of techniques to apply these discoveries to the patient. This is the thrust of medical genetics: the conquest of more and more genetic diseases, either through testing and counseling of parents before pregnancy, or through study of the fetus during pregnancy with the option of an abortion if desired.

In my year of travel I had seen many examples of geneticists at work. I had looked through a high-powered microscope and seen the chromosomes of a genetics clinic patient. To my inexperienced eye they were a jumbled mass of rodlike structures. But to the cytogeneticist standing beside me the genetic error revealed there loomed as large as a truck. Less than a decade ago, the error couldn't have been seen.

In other laboratories I had watched scientists attempting to unravel the mystery of where the genes that control specific human traits are located in the chromosomes. I had observed experiments where researchers attempt to carry this probing of the inner secrets of nature one step further—to actually decode the sequence of chemicals that comprise the elegant and architecturally beautiful double-helix structure of DNA, the substance that is the cornerstone of life. The arrangement of substances along the double helix's spine comprises the genetic code. It is here, in decoding this simple but incredibly versatile system, that the future of genetics awaits humankind. Thoughtful biologists predict that great progress will come from such research in the next two decades. The essence of each person is his or her genetic code. As doctors learn to unravel its most intricate details, there looms the possibility that each person will come to know himself biologically, just as he now knows himself psychically and culturally.

This specter alarms many people, who fear that the ability to know each person in the cold terms of his chemistry will entice evil people to turn such knowledge to selfish gain. All nature of alarmist scenarios can be constructed, from cloning a battalion of new Adolf Hitlers to decreeing that every black couple must undergo prenatal diagnosis and abort any fetus found to carry sickle-cell disease.

Such scenarios run the gamut from the ridiculous to the sublime, with some frightening ones in between. I had encountered enough hints of real problems already at hand to know that these fears can't be dismissed. Hasidic Jews, for example, who arrange the marriages of their young people, are said to favor testing their children for Tay-Sachs disease carriers and arranging marriages so two carriers don't reproduce. On the one hand, wiping out this always fatal disease for which there is no treatment is laudable. But requiring either testing or arranged marriages is odious to many. And what is to be done about parents who seek an amniocentesis to determine the sex of a fetus and then obtain an abortion if the sex doesn't suit their whim?

There is no question that more and more such issues will surface and that doctors, philosophers, theologians, and ordinary citizens will be caught up in the turmoil of how to deal with technology and discovery that are outstripping moral progress. In fact, some doctors are already having to solve these problems on an ad hoc basis with no moral guidelines to rely on other than their personal convictions. I saw one example of this in the neonatal special-care unit at Yale University when the decision was made to turn off the machines and let a severely genetically ill baby die. I saw another at the University of Washington where doctors sought to open sealed adoption records so they could warn the adoptive parents that their daughter might have a genetic problem. In both cases, the doctors handled the problems thoughtfully and ethically. But the potential for a misstep remains.

I found my seat in the auditorium and watched McKusick arranging his notes and conferring with his colleagues. I found myself thinking about something I had read almost a year before written by Lewis Thomas, the doctor at New York's Memorial Sloan-Kettering Cancer Center who has become famous as a thoughtful observer of science and medicine. Reflecting about the role of evolution in the scheme of things, Thomas wrote in the *New England Journal of Medicine*:

> We [mankind] have become so conspicuous that we are in our own light. We seem to have swarmed over everything, elbowed our way into every hidden place, taken over. By every appearance we now look like the most dominant being that ever lived, in total control of the whole operation, and since we know as a deep truth that this cannot be so, it is depressing. Here we are, out in the center of the stage, in full light, pretending that we know how to act the role without understanding the play. It is embarrassing.

Thomas' likening the human being with biological and evolutionary feelings of superiority to someone acting a role without understanding the play had stuck in my mind. I recalled it several times while sitting in the living room of one family or another, listening to them pour out a gripping story of sorrow, disappointment, guilt, and the sheer frustration of living with a genetic disease. In our eagerness to push forward with scientific discovery, the mental and social needs of genetics patients were sometimes being neglected.

A few days before, at a cocktail party in Bar Harbor for journalists attending the short course as observers, I had fallen into conversation with a doctor about the families I had seen. He was fascinated with my "inner" view of the lives of patients, a view he would never see because of the necessity to maintain a professional distance. What was the single thing most common to all the families' problems, he asked.

I had to think a moment to answer the question. What I had discovered, I told him, was that geneticists are among the most sensitive and perceptive of physicians, more concerned about letting their patients make decisions for themselves than any other medical specialists I am acquainted with. As a result, people with genetic problems probably get better medical care than patients visiting any other department in the large medical center.

But what struck me the most, I continued, was that, almost without exception, no one was paying enough attention to helping the families with genetic problems cope with their problem—not in a medical sense but in a psychological sense, in a social sense.

There is Nancy, the little girl in the tiny west Texas town. Her family is having real difficulty accepting their plight. The hostility I encountered there was palpable. In Seattle, Arlene and Charlie are pushing Christine to be as physically active as possible in their own form of denial that something may be wrong with their daughter. Frank and Mary O'Casey in New York City seem to have layer upon layer of feelings trapped inside, unable to let them out, to work them through. Of all the families, perhaps the gospel singers in North Carolina have coped most effectively. They have turned their lives and their destinies over to God completely.

Yes, I told the doctor, if there is one area where medical genetics still often seems to be laboring in the Dark Ages, it is in helping its patients meet their emotional and social problems.

My reflections were interrupted by the beginning of the session. McKusick announced that the first patient would be a boy with Hunter's syndrome, a disease that causes short stature, misshapen bones, and frequently mental retardation.

An assistant opened the door behind the lectern and Raymond came through. After McKusick caught Raymond and steered him to one of the folding chairs, Raymond's mother sat beside him. After a few more remarks, McKusick asked Raymond's mother to tell her story, prompting her at several points with questions.

Raymond quickly became restless as she spoke, describing how the disease had been diagnosed, how her three grown daughters had submitted to tests and learned that they were all carriers of the disease, which is passed along only to male children. Soon Raymond was in his mother's lap. Then he began to seek her attention.

"Mom. Mom. Mom. Mom," he chanted, just as he had done in the lobby.

"Be quiet. I want to talk," she said firmly but kindly.

He continued to call her name and to place his hand over her face. She would brush the hand away. "Mom. Mom. Mom," he called. "Mom, I want to talk."

"You can't talk now. I'm talking."

I began to feel acutely embarrassed for the child and his mother. As in the case of so many of the families I had come to know, I suspected she was subjecting herself and her child to our scrutiny in the hope of contributing something to prevent this scourge from reaching future households. I knew it was a way she had found to ease psychologically some of her burden.

Next came a physical examination of Raymond, in which he removed most of his clothes and marched back and forth in front of the audience. Finally McKusick asked the boy's mother a question I was waiting for. "How has everything worked out for you over the past few years since the diagnosis was made?" he asked.

"Certainly it was a terrible shock when we first learned of it," she answered. "We eventually realized we had to live one day at a time. While it's worrying and upsetting not to know what is going to happen next, sometimes it's better not to project too far into the future."

I am certainly no expert on families with genetic problems, but to me her answer indicated a form of denial. The prospects for a child like Raymond, I knew, were not good. He would probably die in adolescence or early adulthood.

"It's a good thing that all of us don't know what's ahead," McKusick said. The remark drew a laugh from the audience, although I couldn't understand why. The doctor wasn't being flippant.

A few minutes later Raymond and his mother were excused and left the room by the same door they entered. Technical questions from the audience followed. Then, just as McKusick was about to summon the next case, a man on the other side of the room stood up and asked another question. He was short, pudgy, and swarthy and spoke with a thick accent.

"How much is the mother aware of problems that will face Raymond in the years to come?" he asked. "And in what ways will you prepare her for the natural course of the disease?"

McKusick looked startled. "I don't think I understand what you're getting at," he said.

The man repeated his question, making an obvious attempt to pronounce the English words as distinctly as possible.

The question clearly made McKusick uncomfortable. It was a departure from his script. "I agree with her philosophy that you take one day at a time. Usually I urge patients and their responsible family members to take that as their view of life and not try to plan too far in advance."

The questioner was about to interject something more when McKusick added hurriedly, "Perhaps we could delay this until we see the next family, which is another Hunter's syndrome case."

I was astonished by McKusick's answer, however much he had been caught by surprise and answered off the cuff. Is that all there is, I wondered. There must be more to it than that. Looking around, I saw several people shaking their heads. Perhaps they felt as I did. Here, suddenly, was the crux of my experience with families afflicted with errors in their own genetic codes. The man asking the questions had perhaps sensed, as I did, the precision and exactitude of medicine marching on as Raymond was examined and his mother's story recounted. It's easy to deal with the medical facts, the man seemed to be saying with his question, but what about the person sitting there on the chair?

More patients came in. There was another boy with Hunter's syndrome, this child much less severely affected. The family with Charcot-Marie-Tooth disease came through, followed by a teenage boy with Ehlers-Danlos syndrome, a condition of hyperextensibility at the joints and highly elastic but fragile skin. I kept glancing at the man across the room, hoping he would pursue his line of questioning.

Next came a teenage girl with Marfan's syndrome, a disease that causes long limbs, visual problems, and eventual heart and blood vessel difficulties. She ultimately faced possible rupture of the aorta, a major blood vessel, at the heart. The arm span of someone with the condition is often longer than the individual's height.

"What is the arm span in relation to her height?" someone asked as McKusick examined the girl.

"We don't have a tape measure," he said, straightening and smiling. "We have an electron microscope here, but no tape measure." Everyone laughed heartily, including the girl and her mother.

After the patient left the room, the man across the auditorium whom McKusick had cut off rose again. He asked his same question, embroidering it in case there was any doubt about what he wanted to know.

McKusick pursed his lips thoughtfully. "I could make some general points on that," he said. "I think with hereditary diseases in general that you work the parents in on the problem slowly. You don't hit them all at once. You bring along their education in this matter."

The questioner asked what the mother and the daughter should be told about the possibility of the daughter's children having the condition.

"I think the mother should be told this is a condition that can be passed on. We have no prenatal diagnosis for this," McKusick said.

And there ended the attention paid on that particular afternoon to how patients deal with their genetic problems. There were two times, however, when as patients were being dismissed another doctor who helped arrange the session interrupted, saying, "Now that they've been so good to submit to all our questions, let's ask them if they have any questions for us."

Invariably, an avalanche of questions followed.

I left the session feeling depressed. The answers I sought hadn't appeared. The added perspective was missing. What troubled me most was that the initial, acute embarrassment I felt when the first patient and his mother entered had faded as the afternoon wore on and we saw one patient after another. In that short time I felt myself becoming jaded and insensitive to their plight. Perhaps there was my answer. The physician must deal with the medicine first. Maybe there isn't room for much more afterward.

NANCY

David Bilheimer stepped quickly onto the elevator and pushed the button. Even though his mind was on the drama unfolding on the seventh floor, he was self-conscious about his clothes. He had been lounging at home, reading a book as bedtime neared. He looked down at his tattered slacks and the sweat shirt. It certainly wasn't proper dress for a faculty member in the wards of Parkland Hospital or the University of Texas Health Sciences Center in Dallas. But there had been no time to change.

The seventh-floor hall was quiet and the rooms dark because of the late hour. He forced himself to walk more slowly when he entered the research ward—to collect himself. He knew what he would find. He had caught a hint of the anxiety when the nurse called. He must enter the room calm, in control, exuding self-confidence and dispensing assurance to the seven-year-old girl and her mother. He paused outside the door and took a deep breath. Then he went in.

The scene was as he expected. Nancy was propped up in the bed, her round eyes large with fear. She wasn't crying, but he could hear her rapid breathing, see her frail chest rising and falling laboriously.

"Here's Dr. Bilheimer now," Nancy's mother said brightly. The worried creases in her face contradicted her gay tone. She sat on the edge of her daughter's bed in a bathrobe, holding Nancy's hand.

A nurse stood nearby. The resident was out of the room. A portable electrocardiograph had been wheeled up to the bed and the electrodes attached to Nancy's chest.

Nodding to the nurse, smiling at Mrs. Buckingham, David

Bilheimer focused his attention on Nancy. "Have the chest pains come back, Nancy?" he said soothingly. She nodded.

"Do you feel like something is pressing down on your chest?" he asked. She nodded again, rolling her eyes fearfully toward her mother.

Bilheimer took out a stethoscope and bent to listen to the girl's heart. The murmur was still there, caused by aortic stenosis, a constriction where the aorta leaves the heart, which alters the blood flow through a heart valve. But he couldn't detect any new sounds that might reflect a turn for the worse.

He moved the stethoscope away from the heart, listening to the lungs, searching for the rattling sound of rales. That would signal fluid building up in the lungs, indicating the onset of congestive heart failure. But the lungs sounded clear. Watching his wristwatch, he counted her breaths. Her respiration was up: 30 breaths a minute from a normal of about 20. He found her pulse and counted the heartbeats. Her heart rate was greatly increased, up from a normal of 80 to 130. Her blood pressure was up, too. Finally he turned to the electrocardiograph and studied the squiggly lines on the paper tape. The lines reflected the electrical activity of Nancy's heart muscles as her heart pumped blood through her body. The EKG revealed that all the cardiac problems plaguing the little girl were still there, but he couldn't see an indication of a change.

"I'm going to give her some nitroglycerin in addition to the propranolol she's already getting," he said to the nurse. "And I want to get her on oxygen." The nurse left the room for the medication.

"You're going to be okay, Nancy," he said to the girl. "You'll be feeling fine in just a few minutes."

He turned to her mother. Mrs. Buckingham had returned to her bed. "I think it's just another angina attack. She's going to be okay." He smiled reassuringly at her. The nurses had told him that Mrs. Buckingham wasn't sleeping at night. She refused the sleep medication he prescribed. She had confided to him that she lived in constant fear that Nancy would die in the night and that she would be asleep at the moment her daughter needed her the most.

The nitroglycerin and the oxygen and perhaps just David Bilheimer's presence began to work on the little girl. Her breathing quieted. She relaxed. The terrible, frightening pain began to subside. It was after midnight when Bilheimer finally turned his car into the driveway of his home in a quiet suburb of north Dallas.

"Nancy Buckingham was one of the first children to come on the research ward," David Bilheimer told me when I went to Dallas to learn about the landmark discoveries into the mysteries of high cholesterol levels that two of his colleagues have made. In his mid-thirties, Bilheimer is a man of medium height with carefully trimmed brown hair and tastefully chosen clothes who battles potential pudginess by daily workouts at home on an exercise bicycle.

He came to Dallas to join the group formed by Joe Goldstein and Mike Brown, two medical scientists also in their mid-thirties who stunned the genetics world with a series of brilliant discoveries about familial hypercholesterolemia, a hereditary form of high cholesterol that is the most prevalent dominantly inherited genetic condition among Americans, but a problem that most people don't know they have. Its consequence often is early, sudden death from heart disease. Goldstein and Brown persuaded Bilheimer to join their group and assume responsibility for care of the unusual but critically ill patients like Nancy Buckingham who have become the foundation of their research.

"The most important thing on a research ward, the thing you must not lose sight of, is that you must do nothing to harm the patient. This girl had already had a heart attack. I had to ask myself, 'Suppose our manipulating her, putting her through all these tests, causes another heart attack?' I didn't think we were so close to a breakthrough in the research that these tests would have immediate benefits to Nancy. What we were doing would have a payoff four or five years down the road.

"But Nancy's family was in favor of it. They had been through so much and their hope was that our research would eventually provide some new form of therapy for her. I talked to Joe Goldstein about it in detail one day after her angina attacks began, and he thought perhaps we should call it off. But I said, 'Let's wait a few more days and see if she quiets down.' I thought part of it might just be getting used to living in the hospital. The worst attacks seemed to come late at night. And sure enough she quieted down and we went on and completed the study. And what we learned from Nancy proved that Goldstein and Brown's discoveries in the laboratory, in the tissue culture dishes, was actually happening in the patient. In effect, they discovered the cause of familial hypercholesterolemia and Nancy confirmed that they were correct."

It is startling to think a young girl of kindergarten age could have so much cholesterol in her body that her blood vessels are already beginning to clog with fatty deposits, until they resemble those of a person of fifty or sixty. In the second grade Nancy suffered a heart attack and apparently was brought back to life by her mother's cardiac and mouth-to-mouth resuscitation. In the third grade she had open-heart surgery and received artificial valves and new vein grafts in an operation usually performed on people in their forties and fifties who have severe heart disease.

Statistically, there is one Nancy Buckingham for every 1 million people. She has a double dose of the familial hypercholesterolemia gene, a condition that geneticists say is homozygous. As a result, her cholesterol levels are six to ten times higher than normal and her life expectancy is shortened. Cardiac death is the ultimate outcome.

Nancy's parents and two of her three brothers and sisters are what geneticists call heterozygous. They have a single dose of the familial hypercholesterolemia gene. They have cholesterol levels higher than normal, but not nearly so high as Nancy's. Because of this higher fat in their blood, they are much more prone to heart disease in middle age.

I first heard about Nancy when David Bilheimer spoke at a medical meeting. I buttonholed him in the hall afterward and discussed writing about the little girl he had mentioned. Eventually we arranged for me to go to Dallas and then travel on beyond Fort Worth into the rolling mesquite-covered hills of west Texas, to the small town where the Buckinghams live. Nancy's father drives a delivery truck there and her mother is a nurse's aide in a nursing home. Bilheimer likes to see Nancy in his clinic every two or three months, so he arranged for Nancy and her mother to come to Dallas on the day I was there—a three-hour round trip in their aging Pontiac.

Soon after we entered the examining room where Nancy and her mother waited, I found myself listening with a stethoscope to Nancy's heart and gaining an appreciation of David Bilheimer's affection for the little girl, his respect for her mother, and his love of teaching medicine.

Nancy's wide, expressive eyes set in a round face are what you notice first. She is thin, with light brown hair, and wore sneakers, jeans, a T-shirt, and a denim jacket.

She seemed to shrink from Bilheimer's touch at first, although he was exceedingly gentle. But she complied with his instructions to take deep breaths, roll over, sit up, lie down, and stand, willingly, although with a veiled look of resignation. She spoke in a soft voice when she answered his queries about her activities, her medications, and how she felt. When she could, Nancy cast a look at her mother, beseeching her to answer the questions. Mrs. Buckingham did so quickly each time. She is a woman in her mid-thirties who has put on extra pounds over the years but still retains a simple beauty. She hovered nearby, holding Nancy's jacket. It was obvious that a special bond exists between the woman and her youngest child.

Earlier, David Bilheimer had outlined Nancy's medical history in his cramped office, glancing through a file containing her records. She was born normally in May 1967, although her mother quickly noticed strange yellowish bumps on each of her heels. Had Nancy been seen then by a geneticist or cardiologist acquainted with lipid disorders, her condition might have been spotted at once. But the Buckinghams thought nothing of the bumps and their family physician paid them no attention.

In May of 1972, when Nancy was almost five years old, she began fainting, often while playing. The doctors in the small town where the Buckinghams lived could not explain it. The fatty deposits on her heels had spread to the elbows and to her hands between the fingers. When the fainting spells continued and the family physician suggested Nancy's difficulty was psychological, her mother took her to a pediatrician in Fort Worth, the nearest city where specialized medical care could be obtained. This doctor took one look at the yellow bumps, which are deposits of cholesterol under the skin called xanthomas, and diagnosed a cholesterol disorder. Blood tests revealed a cholesterol level of 1000 milligrams percent, five to six times normal. She was referred to a cardiologist, who injected a dye into a blood vessel near the heart and took X rays of the dye's spread through the heart. Her arteries looked like those of a middle-aged or elderly person. Massive fat deposits had built up on the walls. This restricted the blood flow entering the left side of the heart, causing a chain reaction of events that led to fainting, rapid breathing, and chest pains. The high cholesterol was treated with medications, but the level came down only to 800 milligrams percent, still far too high.

The Fort Worth cardiologist had read a medical journal article by Joe Goldstein, who had studied the hereditary forms of high cholesterol while working with geneticist Arno Motulsky at the University of Washington in Seattle. The cardiologist called Goldstein and told him about Nancy. The Dallas physician, who was then thirty-two years old, met the Buckinghams, took blood samples and a medical history, and began constructing their pedigree. The doctors had no doubt that Nancy's condition would worsen.

The fainting spells continued until a particularly frightening one in May of 1974, when Nancy was just seven. She passed out while playing in the yard. Her mother couldn't find a pulse and gave her daughter mouth-to-mouth and cardiac resuscitation until she revived. She was admitted to a Fort Worth hospital for study. The doctors concluded that her heart's function had worsened, perhaps because of a small heart attack. Nancy's mother dealt with the problem by sharply restricting the girl's activities, confining her to the house, and treating her as an invalid.

That July, Mrs. Buckingham and Nancy entered the University of Texas Health Sciences Center in Dallas' research ward at Parkland Hospital for studies to measure in detail their bodies' metabolism of fatty substances. The following October, with her cholesterol levels still dangerously high, Nancy went to Denver, where the noted liver surgeon Thomas Starzl performed an experimental operation to sharply curtail the supply of nutrients in the blood reaching the liver, thus reducing the amount of choles-

terol Nancy's body was manufacturing. She survived the surgery
and her cholesterol level dropped dramatically, to about 400 milligrams percent, still high, but within the range found in ordinary people considered to have high cholesterol. After the liver
surgery Nancy had a growth spurt, but her chest pains, which
were the result of angina, continued. The past damage to her
heart remained and her doctors suspected that even with the
lowered cholesterol levels further heart damage was probably
occurring.

In September of 1975, during a regular examination, Bilheimer
noticed evidence that Nancy's heart might be enlarged and that
a new murmur, involving another valve, had appeared. She was
seen by a pediatric heart surgeon in Forth Worth who confirmed
Bilheimer's suspicions and noted further buildup of fatty deposits on her blood vessel walls. Later that fall she underwent
open-heart surgery in which she was given two artificial heart
valves. Veins taken from her legs were grafted onto the blood
vessels around the heart to improve blood supply to the heart
muscle. It was unusual and difficult surgery in so young a patient. Nancy developed severe chest bleeding afterward and required massive blood transfusions, but she survived.

Her condition improved dramatically after she recovered from
the surgery. The angina caused by the poor blood flow to the
heart muscle was gone, as were the fainting spells. She could
run and play, and even turn cartwheels. She became a nearly
normal member of the family.

A large thick scar runs from just under Nancy's throat to her
abdomen. Stitch mark tracks run up either side of the scar,
attesting to the size of the incision that had to be made to lay
open her chest. An array of sounds filled the stethoscope when I
placed it over her heart as Bilheimer instructed. There was a
loud, distinct clicking sound and a lighter, separate click that
seemed to come from deeper in the heart. These were the valves.
Listening as Bilheimer instructed, I began to distinguish the
sound of the blood rushing past the valves. We listened to one of
the two carotid arteries that rise from the chest into the neck and
on to the brain. There was a rumbling sound in one, what the
doctor called a bruit, caused by turbulent blood flow past one of
the valves. Bilheimer positioned his stethoscope at a point on her

abdomen where I heard the swishing sound of blood flowing past another bruit. He likened each bruit and the sound I heard to water tumbling over the rocks in a babbling brook. Later, he told me this was a sign of the clogging of Nancy's blood vessels by the high cholesterol in her blood.

Nancy has another scar on her side from the surgery to divert blood flow around the liver. There are two scars on the insides of her thighs where blood vessels had been taken for grafting in the heart area.

We listened to her lungs, which were clear, with only the sound of the air rushing in and out. We looked at the half-dozen or so places on her body where the xanthomas had formed. In every case—her heels, elbows, hands, and other points—Bilheimer said the xanthomas had grown smaller, leaving yellowish red areas. Lowering her cholesterol has clearly had beneficial effects. The doctors hope it has been lowered enough to halt clogging of blood vessels and other internal changes. Finally, Bilheimer showed me a yellowish half circle visible in the whites of Nancy's eyes. This, too, is a cholesterol deposit, called an arcus corneae.

"Nancy, you look great," Bilheimer said finally, stepping back to regard the little girl, who sat on the edge of the table. "Do you feel as good as you look?"

She smiled shyly, shrugged her shoulders, and rolled her eyes at her mother. Mrs. Buckingham was beaming at her daughter.

"Nancy never talks," Bilheimer said to me, chuckling. "That's characteristic. She probably has said ten words to me all the time I've known her."

"She's gone from an A for conduct in school to a B because she talks too much," her mother said. "Nancy, you talk too much," she scolded gently.

After we returned to Bilheimer's office, I asked him what had been his most rewarding moment in caring for Nancy.

"To see her come through all of this and return to some sort of a normal life," he said without hesitation. "When she came to us she was confined to the house. She was treated as an invalid. The fact that we were able to lower her cholesterol through the surgery in Denver was very important because we can see the xanthomas going away. We hope the progression of the atherosclerosis will be retarded. The heart surgery has allowed her to go back to school and to be pain-free for a period of time. Those are all rewarding things.

"But it is frustrating in the end because you can't lower the cholesterol low enough in a relatively innocuous way to get her back truly to normality."

"What is the long-term prognosis?" I asked.

Bilheimer studied the titles on the bookshelf above his desk. It was several seconds before he spoke. "It's guarded. If the veins grafted to her heart don't close off and if her heart isn't further damaged, then in a couple more years she will require another operation to give her a larger aortic valve. A small valve was put in because she is a child. She will grow, but the valve won't. So it will become inadequate. Maybe in a couple of years, maybe four or five years."

"You work so hard and invest so much in Nancy for the short term, and yet the long term is dismal," I said. "How do you reconcile that?"

"We don't know what's around the corner," he said. "You struggle to keep someone like Nancy going as best you can. To some extent you have to see how they react to your efforts. If they play out and decide they have had it, you begin to wonder whether your efforts are worthwhile. Are you just succeeding in making the patient miserable? We certainly haven't gotten to that stage in Nancy yet. She is back to enjoying life again and that's rewarding.

"Her life-span will be prolonged because of the aggressive treatment to lower her cholesterol. In the past few years we have witnessed a tremendous increase in the understanding of familial hypercholesterolemia. We now know it's a molecular defect and we know the nature of that defect. We didn't know that in 1973. In just four years we are in a position to begin to design drugs or look for ways that might specifically inhibit cholesterol production in a safe way. It may be that a year from now we will develop a drug which might not only lower Nancy's cholesterol to safe levels but actually reverse the atherosclerosis. You work on that hope. It's what keeps everyone going, including Nancy's family."

The Buckingham family exemplifies a genetic condition inherited in a dominant fashion. There are nearly 900 such dominant genetic diseases.

The genes that control all our genetic traits are located in the chromosomes, structures in the nucleus of each cell. There are 46 chromosomes in humans and they exist in pairs, so geneticists speak of 22 homologous, or similar, chromosome pairs and a pair of sex chromosomes. In the simplest instance, a single gene controls a single genetic trait. This gene is located at a specific point in a particular chromosome. Another gene that controls the same trait is located at the same position on the other member of the chromosome pair. Sometimes the genes on paired chromosomes responsible for a particular trait are identical. Sometimes they differ slightly. This difference is the root of genetic disease.

When sperm are formed in the testes or eggs in the ovaries, the chromosomes undergo division and rearrangement. The 23 pairs of chromosomes divide and separate so that each sperm or egg contains only 23 single chromosomes. One of the genes for a trait is in one sperm or egg and the other in another sperm or egg. When fertilization occurs, 23 chromosomes are contributed by the sperm and 23 chromosomes by the egg to create the first cell of a new individual, containing the correct number of 46 chromosomes. All the thousands and thousands of genes that exist in pairs are reunited. One member of the pair comes from the father, one from the mother.

If the two paired genes are normal, then the baby will be normal. If one or both of the genes are abnormal, that may lead to disease. If only one copy of the abnormal gene is required to cause disease, doctors speak of dominant genetic conditions. If two copies of the abnormal gene are required, doctors speak of recessive conditions. Familial hypercholesterolemia is dominant. That means when the defective gene is present, it takes precedence over the other, normal gene. It can express itself even in the presence of the normal gene.

Half of that parent's sperm or eggs will have the abnormal, dominant gene and half will have the normal gene. As a result, the probability is that half the children born to that parent will have the gene, half will not.

In dominant diseases, the single gene is enough to cause the child to be affected with the disease. The other parent may be perfectly normal, but in dominant genetic diseases it makes no difference.

Sometimes both parents have the dominant defective gene. In that case, half of the eggs at the time of fertilization have the gene and half do not. Likewise, half the sperm have it, half do not. There is a 50 percent chance with each pregnancy that the child will receive one dose of this gene; there is a 25 percent chance that a normal sperm and a normal egg will come together; and a 25 percent chance that an abnormal sperm and an abnormal egg will come together, giving the child a double dose of the gene. In many diseases, this double dose is immediately fatal—if not at the moment of conception, then perhaps sometime during the pregnancy or shortly after birth. In rare instances, the consequences appear later in life.

That was what happened to Nancy Buckingham. She received a double dose, one from each parent, an event that occurs approximately one in every million births in the case of familial hypercholesterolemia.

Nancy's paternal grandfather, Ralph Buckingham, had high cholesterol and died of a heart attack. He was never tested but was presumed to carry a gene for the disease. Her paternal grandmother, who was tested by the Dallas doctors, had normal cholesterol levels. Nancy's father, Harold Buckingham, received the familial hypercholesterolemia gene from his father, as did two other of his four brothers and sisters. All are now in middle age or approaching it, and those with the abnormal gene have high cholesterol levels and run a definite risk of heart disease and possibly sudden death unless they take care to lower their cholesterol.

Nancy's mother, Carolyn Bell, received her familial hypercholesterolemia gene from her father, Leland Bell. Carolyn's mother, Helen Bell, didn't carry the gene. Leland and Helen had four children, and two received the gene.

At the time they married in 1962, Carolyn and Harold Buckingham did not know that each carried a gene for familial hypercholesterolemia. They had four children, of which Nancy is the youngest. The family remained ignorant of the genetic disease until Nancy's difficulties were diagnosed. Subsequent tests revealed that the oldest child, Laura, carries a single dose; Harold, Jr., is free of the condition; and Susan, the third child, also carries a single dose.

"Harold and I never would have had children if we had known about this," Carolyn Buckingham told me wistfully as we chatted in the hall of Parkland Hospital that day while Nancy was still inside the examining room being weighed and tested by David Bilheimer.

She stared down the hall without speaking. "Maybe we wouldn't even have married if we'd known," she said finally.

Joe Goldstein was the first out-of-state student to attend the University of Texas Southwestern Medical School. By several accounts, he was also one of the brightest medical students trained there in many years. When he graduated in 1966, so the story goes, Goldstein was told that after he finished his postgraduate training, including a period studying genetics, the medical school administration might entertain the idea of inviting him back to Dallas to create and head a new division of medical genetics within the department of internal medicine.

After graduation, Goldstein, who was born in Sumter, South Carolina, went to the Massachusetts General Hospital in Boston, where he was an intern and then assistant resident in medicine.

Mike Brown, who was born in New York City, went to the University of Pennsylvania as an undergraduate and then on to the university's medical school, graduating the same year as Goldstein. He met Goldstein at Massachusetts General, where he also was first an intern and then a resident in medicine.

David Bilheimer, who was a year behind the other two, was an undergraduate at Muhlenburg College in Pennsylvania. He and Brown belonged to the same medical school fraternity at the University of Pennsylvania and became friends. During his last year in medical school, Bilheimer was considering going to Massachusetts General as an intern and visited Mike Brown there. Brown introduced him to Joe Goldstein.

The following year, Bilheimer and Goldstein were teamed up at Mass General—Goldstein was the assistant medical resident,

Bilheimer his intern. "Goldstein ran the ward and I ran after him," Bilheimer says.

This was in the late 1960s, when all young physicians were obligated for some type of military service. Goldstein and Brown and then Bilheimer, a year later, all secured positions at the National Institutes of Health in Bethesda, Maryland. Goldstein spent two years in a laboratory at the National Heart and Lung Institute and then joined Arno Motulsky, a nationally known geneticist at the University of Washington in Seattle, for a two-year fellowship. Brown spent two years in a laboratory at the National Institute of Arthritis and Metabolic Diseases and a third year in Bethesda at the National Heart and Lung Institute. Bilheimer spent four years in a laboratory at the National Heart and Lung Institute.

During their association, Goldstein repeatedly extolled the virtues of the medical school in Dallas to his two colleagues. While Goldstein was completing his second year in Seattle, Brown joined the Dallas faculty as an instructor in the department of internal medicine. A year later, in 1972, Goldstein returned to his alma mater as an assistant professor and head of the new division of medical genetics. In 1973 Goldstein and Brown persuaded Bilheimer to come to Dallas as an assistant professor to handle the clinical care of their research patients and conduct his own research into the biochemistry of lipids. By then, Goldstein and Brown were already becoming famous for their research into the genetics and biochemistry of familial hypercholesterolemia. Within four years, the two doctors would each have published more than one hundred scientific articles in medical journals and risen to the rank of full professor with an endowed chair in the department of medicine, all before age forty.

In the ordinary course of medical research, several research teams across the country or around the world are at work on the same problem. A small finding by one group is used by another to make a further step in the solution of a problem, which will then be used by yet another group. But occasionally someone makes such a fundamental discovery that he or she suddenly leaps far ahead of the field and races forward, the answers to some medical puzzle tumbling forth with lightning rapidity.

Sometimes within the space of two or three years such a scientist may have his name attached to dozens, even hundreds, of research papers. Major prizes are awarded, academic careers rocket upward, and other major universities attempt to entice the scientist away.

David Bilheimer was working in laboratories at the National Heart and Lung Institute in Bethesda, concentrating on the biochemistry of lipids, when one of Goldstein and Brown's first papers appeared. He remembers the instant sensation it caused.

Goldstein and Brown had taken skin cells from a child homozygous for familial hypercholesterolemia, grown them in a culture dish, and shown there was a defect in the way the cells dealt with cholesterol. The elegance of their work was stunning.

Many chemical reactions move forward only in the presence of enzymes, which are special proteins. The two scientists had developed a method to test for the presence of the cholesterol-forming enzyme in their skin cell cultures. They studied how the enzyme's presence varied in relation to cholesterol. The research confirmed that if the normal cells were bathed in nutrients rich in cholesterol-containing substances, the cells' internal cholesterol-forming machinery was turned off. If the nutrients were devoid of cholesterol, the cells' internal cholesterol factories started up.

When they performed the same test with cells taken from a familial hypercholesterolemia homozygote, they made a striking discovery. It didn't matter whether cholesterol substances were present in the culture nutrients or not. The cells produced cholesterol internally at a maddening pace.

That finding led to a rapid series of discoveries by Goldstein and Brown. More scientific papers about the cause of familial hypercholesterolemia came like an avalanche.

All cells have membranes, a complex cover that surrounds the cell and controls the substances that enter or leave it. Goldstein and Brown discovered that the cause of the hypercholesterolemia in their homozygous patients was an almost total lack of receptors on the membranes that would permit cholesterol-containing substances in the cell's environment to enter the cell. With no external source of cholesterol available, the cells' internal factories produced it nonstop, unknowingly flooding the body with too much of the fatty material. The signaling system present in normal people, involving the receptors on the cell membranes, was missing in the homozygotes. In the heterozygotes—those with a single dose of the gene—the number of functioning receptors was diminished by about half, accounting for their cholesterol levels, which are about halfway between normal and the extreme found in people with a double dose of the gene.

Pursuing studies of the receptor defect further by including more homozygotes, Goldstein and Brown then found there are distinctly different forms of genetically transmitted receptor defects. The first form they found had no receptor sites. The second form they found had receptor sites, but only about 10 percent of them functioned. The third form they found had receptor sites, but a defect was present in the internal machinery manufacturing cholesterol. They demonstrated that these different deficiencies of cholesterol metabolism are different genetically.

Finally they suggested that if a person has two normal genes, cholesterol metabolism is normal. If that person has one normal gene and one of the three possible defective genes, he or she will be a heterozygote for that particular form of familial hypercholesterolemia. If there are two of the defective genes, he or she will be a homozygote with one of the three possible forms.

It is one thing to work out such things in a culture dish in the precisely controlled environment of the laboratory. It is a very different thing to prove that it is also occurring in the patient. Not long after David Bilheimer arrived in Dallas in 1973, he made plans to attempt to validate the laboratory findings with Nancy and her mother, and later with one of Nancy's sisters and another patient.

When Nancy and her mother were admitted to the research ward for their six-week stay, complicated in the beginning by Nancy's angina attacks, they were fed a carefully controlled low-fat diet. Bilheimer and his associates went to great lengths to make it palatable, resorting to such things as blending it or freezing it to resemble ice cream. By controlling how much fat Nancy and her mother received in their diets during the research period and making daily measurements of the levels of fatty substances in the blood, Bilheimer was able to monitor what was occurring in the millions and millions of cells in their bodies.

As he had hoped, the results agreed with the cell culture findings. Regardless of how much cholesterol Nancy had in her blood, her body continued to make large amounts of it. The control mechanism that should sense large cholesterol buildups in the blood and shut off manufacture of the material within the cells was missing. This finding was a fundamental contribution to medicine. No matter how rapidly Goldstein and Brown moved forward in their tissue culture studies, their discoveries would ultimately be useful only if they related to actual patients.

It is here that future research may reveal ways to help patients like Nancy, even if it comes too late for her. Perhaps defective receptor sites can be activated. Or, since doctors know in intimate detail all the steps in the cholesterol system, it may be possible to intervene at some point with a new drug to rid the body of large amounts of cholesterol. This is the hope that keeps David Bilheimer going, the hope that Harold and Carolyn Buckingham cling to in the face of the knowledge that most familial hypercholesterolemia homozygotes like their youngest daughter die of heart attacks before age twenty.

When one reads the scientific papers that set out a series of brilliant discoveries, they seem deceptively simple in hindsight. But there must be one spark of genius, one profound insight. Certainly in Goldstein and Brown's case, one was the idea to grow skin cells in culture. Another was the interpretation of the data itself. Many scientific projects falter at this point. Data is discarded or ignored simply because the scientist fails to look at it in an original way, is unable to make the leap away from letting preconceptions dictate reactions. Goldstein and Brown somehow avoided that. Their discoveries were very important and ensured them careers at the peak of the medical research establishment.

It seems an unwritten rule of medical research that the investigator must avoid getting to know his patient too well, becoming too involved in the case.

At one point during our discussions, I asked David Bilheimer to what extent he had become emotionally involved in Nancy's case; what the effect of her death might be.

"I would not be unaffected by her death," he said slowly, ". . . after all that she has been through."

I suspected that he also meant he would not be unaffected because of all *they* had been through together.

From the medical school in Dallas, I was to go to the town where the Buckinghams lived.

"You must tell me what their life is like," Bilheimer said. I wondered if he was envious. He had never been in their home or seen their family life. But if he was envious, he also might decline to trade places with me if invited. There might be such a thing as knowing too much about a patient.

It didn't occur to me that such a philosophy might apply to me. I was intrigued by the prospect of visiting the Buckinghams. I wanted to gauge the effect the burden of a genetic disease had on the family's views toward Nancy and the disease itself. One of the most perplexing problems that faces medical geneticists is persuading some of their patients to take a problem seriously. Familial hypercholesterolemia is a classic example.

Heterozygotes can live into their thirties and forties or beyond and never feel the effects of their disease. They can eat indiscriminately, smoke, drink, and do all the things that raise the risk of heart disease in even genetically normal people. Then one day without warning they collapse and die of a heart attack. Bilheimer spoke of the difficulty of persuading such patients with the genetic disease that they have a problem. A few months before, a member of a large familial hypercholesterolemia family had died suddenly of a heart attack, causing a renewed flurry of interest on the part of other family members in diet, medications, and life-style. But, from experience, he knew that with time they would begin to backslide again.

In Nancy's case, I wanted to know how the family members were coping with their own genetic problem.

I drove west from Dallas on Saturday morning, admiring an utterly cloudless sky and drinking in the limitless vistas of the Southwest. The area beyond Fort Worth consists of rocky, rolling hills covered with mesquite bushes and prickly pear cactus. The people make their livings from small farms, cattle ranches, and scattered areas where oil and gas are pumped from the ground.

The area is dotted with small towns whose names reflect the pioneers of a century ago and their concerns: Brownwood, Big Spring, Comanche, Midland, Big Lake, and Shallowater. Each town is presided over by a round water tower, and the Buckinghams' town, which I'll give the fictitious name of Dalton, was no exception.

Dalton has about 3,000 people. The men wear pointed boots and straw cowboy hats with the crowns and brims creased at sharp angles. They chew tobacco and spit it in pop bottles or coffee cans that they carry in their pickup trucks. Every vehicle has a rack in the back window from which one or two rifles hang. There are a few stores along a dusty main street, a high school, and the nursing home where Mrs. Buckingham works. Dalton's single establishment offering overnight accommodations is a "tourist court" from an earlier era. Each cabin stands apart and has a corrugated metal roof awning under which a car can be parked. I rented one of these rooms for $9.50 and surveyed its accoutrements: a narrow, sagging bed and a sheet-metal shower stall. The room was heated by an open-flame gas heater that wasn't vented to the outside.

I found the Buckingham house on a narrow, paved road on the edge of Dalton among a smattering of houses set on half-acre lots. There are no lawns or landscaping. Hulks of old automobiles rust in many front yards, and old barns seem in danger of falling. Collections of cattle, goats, pigs, sheep, and barking dogs roam about. The family lives in a small, single-story, weather-beaten clapboard house approached via a rocky driveway. I could tell from the state of the roof that water would drip into the rooms when the occasional rains came.

Four children were playing kickball in the driveway when I drove up at 4 P.M. Mrs. Buckingham had gotten off her shift at the nursing home thirty minutes earlier. I immediately recognized Nancy, but had the distinct impression the other children were ignoring me. Nancy didn't speak, but I saw a tiny, secretive wave of recognition as I climbed the steps to the porch.

Mrs. Buckingham received me shyly and led me into the kitchen, where we sat down at the table to talk. Passing through the living room, I saw peeling paint and heavily worn furniture. A brightly lit aquarium filled with tropical fish was in one corner. The linoleum floor in the kitchen was worn bare and cracked, and some of the kitchen chairs seemed to provide a precarious perch at best. But I saw at once that the house was neat and spotlessly clean.

A few minutes after I arrived, Harold Buckingham drove up in his delivery truck, having finished his Saturday rounds. He is a tall, florid man who smokes heavily and is about thirty pounds overweight, with a bulging beer belly. He has striking silver hair combed back in waves and held in place with pomade. He must have been handsome as a young man, but clearly the years have worn upon him. He greeted me, took a beer from the refrigerator, apologized for the cheapness of the brand, and then went into the living room, where he removed his shirt and shoes and socks and sat down to watch a TV sports program. I turned on my tape recorder and Carolyn Buckingham told me about life with Nancy.

"Nobody was ever honest with us and told us straight out that Nancy will die of this until after she had her heart attack in 1974," she said. "We were living near Brownwood then and I was working in a greenhouse. I took the kids to work with me and they were playing outside. We came home for lunch and one of the kids came running in and said Nancy was sick. I ran out on the porch and she was sort of stumbling up the steps. And then she fell unconscious.

"The doctors had told me when she fainted to put her on the couch and wait for her to come to. So I did that. But she didn't come to. And then I couldn't find a pulse and I wasn't sure if she was breathing. I became frantic and gave her mouth-to-mouth resuscitation. I didn't know how to do cardiac resuscitation, but I tried it anyway. Finally, she came to.

"I got her in the car and we drove ninety miles an hour to the hospital in Brownwood. They gave her oxygen and talked by telephone with the cardiologist in Fort Worth. He wanted her to come to Children's Hospital by ambulance the next day. But later they decided she could come by car, so I drove her. She stayed in Children's three days and then came home. She had a low blood count and they said she wasn't getting enough oxygen and that's probably why the fainting was so long. Later I think they decided she had had a heart attack.

"The cardiologist in Fort Worth was very much against heart surgery, although that was what she needed. He said because her cholesterol was so very high that the surgery wouldn't do any good. Everything would just plug up again. That was before the liver bypass surgery, which brought her cholesterol down to levels that they can do something about.

"When Nancy was ready to leave the hospital, that was when the doctors told me for the first time that she would not live much longer. The cardiologist had been avoiding telling us this all the time we had been seeing him. It was quite a blow to me. Of course I didn't tell Nancy, but we immediately cut back her playing. We had already cut it back, but we cut it all the way back. She was almost bedridden."

When Harold first walked into the kitchen, I invited him to join us. He declined and went to the television. But as we talked, he periodically rose from the worn recliner and stood in the kitchen door to listen. It made Carolyn nervous. At one point he came and sat at the table for two or three minutes. We were discussing the genetics of familial hypercholesterolemia and I was trying to explain some of the details of dominant inheritance.

"This cholesterol business is all in what you eat, all in your ingestion," Harold said abruptly, rising from the table. His voice was filled with disgust. "All ingestion," he said as he returned to his sports program.

Carolyn and I continued our conversation. I could tell she was irritated with her husband, but she didn't say anything. He came to the door again a few minutes later. "It's vitamins that's the answer," he said. "These doctors don't pay enough attention to the vitamins we should be taking."

Later, he returned to the table and the conversation turned to what living in New York is like. Both Harold and Carolyn launched into a tirade about the Rockefeller family, arguing earnestly that the family controls everything that happens in Washington. Jimmy Carter couldn't have been elected without their permission, they told me. I steered the conversation back to their daughter. Harold returned to his television. I asked Carolyn to tell me about Nancy's open-heart surgery, which had taken place a year before.

"Dr. Bilheimer came to Fort Worth to be in the operating room with Nancy. They were going to replace her valves with artificial ones and he was going to take her valves back to Dallas to study. Nancy was determined that Dr. Bilheimer be in the operating room. She said she wouldn't go unless he was there with her. He really is her favorite doctor.

"About two hours before she was to go into the operating room Nancy began to be afraid she was going to die. She cried and begged us not to make her go. Back when Nancy was younger, maybe about six or seven, we had to tell her the facts. She was a very difficult patient. It took four or five people to hold her down just to take a blood sample. So the cardiologist sat her down once and told her that if she didn't cooperate she might die.

"They let all our kids come into the room with her and some of our relatives were there. Then it all just came out. She got the idea that if everyone was there, then she really was going to die. I kept saying, 'Nancy, that's not true.' She wanted everyone to go away. So we asked them to leave.

"Dr. Bilheimer was in the operating room, and I've always considered him a doctor who would tell me the truth. When he came out grinning, I knew she had come through. Well, I'm always the type to say, 'Okay, what next? Okay, she's out of surgery, what next?' About two hours after they brought her into the intensive-care unit, she began to bleed badly. So they took her back to surgery and opened her up and tried to stop the bleeding. But it wouldn't stop. It bled all night. The next day they took her into surgery a third time. By this time I couldn't take any more. They gave me a Valium and put us in a little family room they have. I know now they put you there when they don't expect someone to live.

"I woke up from sleeping off the Valium just as they were bringing her back from surgery the third time. The surgeon told us he had opened her up and suctioned it out but there wasn't much he could do. It was bleeding everywhere. He said if it didn't stop in a few hours she would be gone. Harold told me later that he went into the ICU once and her heart was beating so fast they thought she was about gone.

"Our preacher came up that afternoon and we went to the hospital chapel. We did everything. We said verses over her. I even bargained with God. My sister went in the ICU at three and they said Nancy was still bleeding. If it didn't stop in a few minutes she would be gone. She went in at four and the bleeding had stopped. They didn't know why. It had just stopped. Our prayers had been answered. They had given her more than sixty pints of blood.

"It was almost two weeks before I could really talk to Nancy. One of the first things she asked was whether Dr. Bilheimer had been in the operating room."

Nancy and her brother and two sisters came in from playing then, and Mrs. Buckingham suggested they get out their Monopoly game and play with me. Only Nancy seemed to approach the idea with any enthusiasm.

The others didn't want to play Monopoly, so we settled on a game called Yahtzee, which involves rolling several dice. I had never played it before, and Nancy sat next to me and coached me as she played. She turned out to be a warm and talkative child. But even above the noise of the game I could hear the louder of the two artificial valves in her heart clicking each time it snapped shut.

I was only half following the Yahtzee game, concentrating on Nancy, when I became aware that the other children were snickering about something and waiting for my reaction. I looked around the table. "What's so funny?" I asked. No one said anything. The oldest girl laughed out loud and then Harold, Jr., snickered again.

"They're cheating," Nancy said. "They've been cheating you all along."

I felt a sudden flash of anger. Nancy's siblings were displaying their resentment of my presence. Or were they? Perhaps what they were really revealing was their resentment of their little sister. I hadn't fully considered the idea that the girl's brother and sisters and even her parents might harbor deep inner feelings of resentment of the way her illness had altered their lives. I wondered if the undercurrent of hostility I felt their father was

exhibiting toward me also reflected resentment of his daughter.
"Cheating isn't very nice," I said lamely to the children, at a loss
for words.

Nancy's mother appeared and rescued me. She sent Nancy off
for a bath. Harold, Jr., went to his room and the two other girls
disappeared into the other bedroom. I wandered into the living
room, where Harold still sat with bare chest and feet, drinking a
beer and watching television.

There was an uneasy silence for a few minutes before he
spoke. "Do you really think this cholesterol business is something in the genes?" he asked abruptly.

The question stunned me. After all he had been through, how
could he doubt that? The disease and the genetics involved had
been explained to him many times by the Dallas doctors and
others. I knew I was seeing a classic case of the denial that geneticists sometimes encounter. Patients or the parents of patients
simply refuse to believe that anything is amiss in their genetic
makeup.

Harold turned the conversation to money and the difficulty he
had making ends meet. He told me of the succession of jobs he
had held over the years, often as a laborer or delivery truck
driver, and of the different towns in which his family had
lived.

"When Carolyn and Nancy were in Denver for that surgery,
we farmed the other kids out all over the country to relatives. I
couldn't keep them and work at the same time. I had so damn
many bills to pay you wouldn't believe. One night I had all the
bills spread out on the table, trying to figure out which ones I
could pay something on and which ones would have to wait.

"Damned if this bill collector for a finance company didn't come
to the door. So I told him my story. And I invited him in to see
all these bills. He looked them over real careful and said, 'Mister,
I've heard a lot of hard-luck stories in my time, but by God, you
do have problems.' We worked out something where I'd pay so
much a week and pay off their loan."

He talked about the cost of Nancy's care. The Denver surgery
had been paid out of research funds. All of the care at the medical school in Dallas had been free. A program for crippled children run by the state of Texas had paid for the open-heart sur-

gery, which had cost $40,000. He estimated that Nancy's illness had cost more than $100,000, although most of this the family hadn't been burdened with. But the obligations they did have were still staggering for their limited resources. They had to pay for her periodic blood tests and much of the half-dozen medications she had to take each day. Taking time off from work to ferry Nancy for tests and checkups and the like cost them lost income. "These doctors think we can just drop everything and run to Dallas every time the whim strikes them," he said.

There was no mistaking the resentment and feelings of emasculation his voice reflected. I wondered how many family members had wished at some point that Nancy would die. If they hadn't consciously admitted to such feelings, were the feelings yet lurking in the subconscious? Counseling, even if a man as obviously prideful as Harold Buckingham were to agree to it, wouldn't be available. They couldn't afford it. If it were free, they couldn't afford the time involved to travel to Fort Worth or Dallas to obtain it. I began to feel depressed by everything I was seeing.

I had told Mrs. Buckingham during our conversation in Dallas that I would spend Saturday evening with the family and then return Sunday afternoon. I had visions of taking the children on an outing or otherwise spending some time with them. Now I began to wonder if I could back out of my commitment. I knew I couldn't bear to return the next day.

Nancy appeared in a robe and sat beside me. She was warming to me by the minute. She dragged out school books and displayed her assignments. She exhibited phonograph records. The country and western vocalist Olivia Newton-John was her favorite. She sang snatches of verse from one of the singer's songs popular then. At 9 P.M. her mother appeared and announced Nancy's bedtime.

I said my good-byes to Nancy and then told her parents I had gathered enough material and didn't need to return the next day. I shook hands all around, expressed my thanks, fled to the car, and returned to the motel. The Buckinghams were depressing, their neighborhood was depressing, and so was Dalton, Texas.

I had reservations on a flight leaving Dallas late the next afternoon. But I didn't want to stay in Dalton one minute longer than necessary. Using the telephone booth outside the motel office, I arranged to leave Dallas at 5 A.M. the next day. Flights were spotty on a Sunday morning, but I could be back in New York by midday.

Sleeping in closed rooms with an unvented gas heater is dangerous. So I opened the room's two windows wide, turned the heater on high, and climbed into bed to await three-thirty. When the alarm went off, I loaded my suitcase into the car and headed out of Dalton toward Fort Worth and the airport. My route took me past the turnoff for the road leading to the Buckinghams' house.

An impulse made me turn onto the road. I drove slowly down the narrow track and coasted past the Buckingham house. It was dark. Not a light shone anywhere within. I drove down the road, turned around, and coasted by the house again.

I knew the window of the room where the three girls slept. I studied it as I passed the second time, imagining the stillness within, the sounds of three people breathing lightly in sleep. I knew if I stood in the door of the room perfectly still and listened very carefully, I would be able to hear Nancy's heart ticking. The more I concentrated on the sound, the more it would swell to fill the room.

MARSHA AND GARY

How much despair can a young couple endure when they receive their baby's death sentence? When their intense anger and bitterness surfaces, at whom can they direct it? Their doctors? The rabbi and the elders of their synagogue? Their parents? Themselves? The cruelest truth of all comes later: if each had married someone else, they would have been spared.

Harold Nitowsky, a geneticist at the Albert Einstein College of Medicine in the Bronx in New York City who is an expert in Tay-Sachs disease, the "Jewish" disease, begins this story. He is a quiet, seemingly shy physician who is regarded with reverence and awe by most of the couples who come to his office. He sits behind his desk, hands folded on top, the intensity of his feelings occasionally peeking through from behind his quiet manner.

"Marsha and Gary's case goes back to October 6, 1975. I remember it vividly," he begins. "It was a Monday. The preceding day we were involved in a screening program for Tay-Sachs disease at the Central Synagogue on Lexington Avenue in Manhattan. We have screened approximately nine thousand people in the New York metropolitan area, and we have had screenings in Albany, Syracuse, Stonybrook, and generally covered much of the state. Tay-Sachs carriers—people who have the Tay-Sachs gene—are ten times more prevalent among the Jewish population who have their ancestral origins in central and eastern Europe. These are the Ashkenazi Jews. Among couples of this background the disease is present in about one in every 3,000 births. Among the non-Jewish population, it is much lower, perhaps one in every 300,000 births. So we wanted to attempt mass screening in this Jewish population. And that's what we were doing at that synagogue on that Sunday.

"Tay-Sachs has an early onset and leads to a rapid deterioration of the central nervous system. Tay-Sachs children lose contact with their environment and show regression in whatever capabilities they develop in terms of motor development. They eventually have to be tube-fed. They become spastic and have convulsions. The end of their disease is a vegetative state. The cause of death is usually an infection or pneumonia. They become debilitated because it is very hard to maintain their nutrition. These children have been known to live up to five years of age, but many die at age two or three. However, Tay-Sachs is a genetic disease whose carriers can be easily screened.

"We received this call from a young man. He had heard about the screening and called the synagogue and asked to speak to one of the doctors. He was very, very distraught. He was crying, obviously terribly upset. He said his daughter had just been diagnosed as having Tay-Sachs. He asked if we could see her as soon as possible, so we made arrangements for him and his wife to bring their daughter to our offices first thing the next morning.

"Marsha and Gary had been married for five years. This was their first child, born eleven months earlier after a planned, uncomplicated pregnancy. She was apparently doing very well and even recognizing faces, rolling over and smiling and so on. But she was never able to sit up unsupported. At about four or five months of age she was noted to have a startle reaction to loud noise, which is an early clinical sign.

"Marsha and Gary began expressing some concern to their pediatrician and he allayed their fears, saying things seemed to be going along okay. Of course, you never know what patients' perceptions are. But during the course of the next several months they noted that the pediatrician was often checking the baby's eyes with an ophthalmoscope. During that time the baby began showing less attentiveness. She rolled over less frequently and became less responsive. They really became quite concerned. They were eventually referred to an ophthalmologist, because the pediatrician must have spotted something in the fundus of the eye. He probably saw the cherry-red spot. The ophthalmologist found the spot. On the basis of their Jewish ancestry and the infant's medical history, he made a presumptive diagnosis of Tay-Sachs disease.

"It's ironic that they are residents of Queens and live in a neighborhood where we had a Tay-Sachs carrier testing program eighteen months earlier, in April of 1974. They never heard of the testing. So this shows you how when you mount what you think is a very good education program for the community, you still are not reaching all the people, particularly many of the young people whom you need to reach most. We had announcements on the radio and in the local press, but they weren't attuned to it. People hear what they want to hear. So in a sense this could have been prevented if we had gotten to them then. We could have detected the fact they were both carriers of the Tay-Sachs gene with a simple blood test."

Marsha and Gary are a handsome young couple who live in a single-family house they've remodeled on a small plot in Queens a few blocks from the Utopia Parkway. It is a neighborhood of tree-shaded streets and old single- and double-family houses. It could be any American suburb, although finding it in Queens is startling at first if you equate New York with massive skyscrapers and towering apartment buildings. Gary's family business manufactures cardboard boxes in a plant nearby on Long Island. He works there with his father and a brother. Marsha is a financial officer with an air freight firm at one of New York's airports. I spoke with them in their home and quickly realized the intensity of the emotion that talking about their ordeal unleashed. At first they both talked at once, constantly interrupting one another, their words pouring out almost uncontrollably. As we talked they seemed to become calmer, as if the opportunity to tell their story to an outsider was a cathartic experience, a form of therapy.

"I honestly had never ever heard of Tay-Sachs until Jennifer was eight months old," Marsha began. It was a hot summer day and she wore shorts and sandals. She sat cross-legged on the couch in their den. Gary sat at the other end of the couch, sipping a beer. He had come in from sunbathing in the small backyard when I arrived.

"We were in Florida and staying with close friends. One night a friend of my girl friend came over and the topic came up very casually. This girl was talking about this disease because she had received a telegram saying that her sister had been found to be a Tay-Sachs carrier. Therefore her sister was telling her to go and be tested. She had gone to a doctor and he couldn't even tell her where to go for this testing. She was griping about how incompetent these doctors were. I said, 'What are you talking about?' She said, 'Tay-Sachs disease.' It was a genetic disease mostly among Jewish couples. It was fatal and the babies often didn't live past age five. That was all I heard. My daughter at this point was eight months old. I didn't make the connection at all.

"But we still felt something was wrong. I think Gary blocked it out more than me. From the time she was about eight and a half months old I became convinced something was wrong. My pediatrician would say that she was slow but she will catch up. I guess this must have been bothering me when I was in Florida because I made an appointment to take Jennifer to my girl friend's pediatrician. I went on the pretense that her ear was bothering her. He examined her and said, 'Are you happy with your daughter's progress?' I told him no. He said he could do a quick test, which he did, and said her motor skills were at the four-month-old level. He said it could mean nothing, but if I was going to be in Florida for a while he could arrange a neurological examination. But we were coming back to New York, so he said I should speak to my pediatrician in New York. When I did, my New York pediatrician just said all the doctors in Florida are neurotic. They test for everything."

Gary picked up the story.

"We were going to the same pediatrician who had cared for me as a child. I had a lot of confidence in him. I felt I knew him well. But we were new parents. We didn't know the progression that a baby makes as it grows. We did realize she was very slow. But it turned out that she developed more than most Tay-Sachs babies. She progressed further than most.

"When she was eleven months old her eyes started to cross. Marsha was still complaining that something was wrong and I was getting worried, too. Marsha was ill, so I took Jennifer to the pediatrician. I told him our concerns and he suggested we get her eyes checked. He sent us to an ophthalmologist. We went a couple of weeks later and the ophthalmologist saw the cherry-red spot and diagnosed it immediately."

Marsha interrupted.

"You see, I was working full-time. I had someone in the house during the day. But I was ready to quit my job. I felt that I was failing Jenny by working. I thought I didn't give Jenny enough time to teach her things. I felt guilty. I was convincing myself that I was at fault. But Gary said I shouldn't quit. He said, 'You can't teach her to crawl and sit up.'

"Gary was very confident up to this point because the pediatrician was a doctor who had been his doctor as a boy. Gary felt he was really qualified. But the ophthalmologist did a quick eye exam and said, 'I see a grave situation. I think you should go back and see your pediatrician.'

"Well, my God. I was about to explode. I cornered him and said, 'What do you mean, a grave situation?' I wasn't leaving until he told me. So he said he was pretty sure it was Tay-Sachs, a 90 percent chance. We left his office and went right back to the pediatrician's office and he just looked at us as if he couldn't do anything. And then we knew that he had known all along what it was. He looked at Jenny there in his office and she was smiling and moving around and he said, 'I wasn't sure because she was responding up to a certain point, more than the usual Tay-Sachs baby.' He made it clear that basically he didn't want to handle Jenny anymore. He told us to contact the Tay-Sachs Association."

"My parents had been concerned, too," Gary said. "They felt Jenny wasn't progressing, either. They wanted to know the results of the eye examination as soon as we knew. They had gone away for the weekend. I called them and they drove back home and came to our house on Sunday. We were just sitting around the den feeling terribly depressed, and my dad suddenly said he remembered seeing a notice about a testing program in some synagogue. It was on a bulletin board or something at this synagogue. He called there and found out where the testing was and then called that synagogue. It happened to be in Manhattan. He spoke to some of the Einstein College of Medicine people and then put me on the phone. I was really a wreck by that time, not making much sense. But they made an appointment for us next day. We were there at 9 A.M. with Jenny. That's when we met Dr. Nitowsky and Sandy Silverman and the others. They are fantastic people."

Sandy Silverman was middle-aged when she embarked on a second career as a genetics counselor after raising a family. Earlier, she spent many years in laboratory genetics research, studying under Joshua Lederberg, who would later earn a Nobel prize. After her children were grown, she went to a special program at Sarah Lawrence College in Bronxville, New York, to study to become a genetics counselor. She would work with patients rather than on laboratory experiments. One of the lecturers in the program was Harold Nitowsky. When she finished the program in 1972, she joined Nitowsky's staff at Albert Einstein.

"Tay-Sachs disease is a particularly brutal one," she told me. "I've been in the room when couples have had that terrible death sentence pronounced for their child. It's horrible for everyone involved—the person giving the sentence and those receiving it. Very often, even though other physicians are fairly sure of the diagnosis, they will push off on the geneticist the problem of telling the family.

"I've learned to watch the couple when they are told. They do one of two things. They either move together, physically reach for each other. Or they move apart. One man once jumped up and ran out of the room. Gary and Marsha came together. They reached out and touched each other. I saw this happen again and again during their ordeal. When one was down, the other was strong. Marsha was strong during the early period while Gary was nearly going to pieces. Later, when Marsha was down, Gary was strong.

"The first day they came to us, Dr. Nitowsky drew blood for laboratory analysis. When he examined the baby's eyes he could see the cherry-red spot. However, there are some other diseases that masquerade as Tay-Sachs. Dr. Nitowsky laid it out and he was not optimistic. He said if the blood tests revealed that it was Tay-Sachs they were fortunate in one respect. There is a prenatal test for Tay-Sachs. That meant Marsha could have this test during the second trimester of each pregnancy and have an abortion if the baby was affected. They would be able to have normal children. When a couple finish with Dr. Nitowsky and come out of his office, I kind of hover over them. They can't get rid of the emotions they're feeling in the doctor's office situation. So Gary and Marsha came into my office and sat down and cried. And we talked. And after that, Marsha just started calling me. Gary was really having almost a physical breakdown and there was a time when she called me every day.

"Like so many people in this situation, they needed a neutral party to talk to. Friends or relatives don't understand the problem and they often talk in reassuring platitudes of hope when there is no hope. That can be more cruel in the end. Marsha needed a neutral party, someone who knows that her child is going to die and won't offer any false hope for cures.

"There is something you must understand about a recessive genetic disease. It's subtle. But eventually it occurs to most couples like Marsha and Gary: if their baby has a recessive condition, then that means each of them could have a normal baby with somebody else, with almost anybody else, in fact. When this realization comes, it challenges the relationship. I've seen marriages that became very shaky. I've seen couples separate. There is enormous guilt. It's a difficult test of a relationship."

Each of us has from six to eight recessive, deleterious genes. Most people can live a lifetime and never know they exist. They are revealed only if we marry someone carrying the same recessive gene and one of the children receives a double dose of the gene. That child will be affected with the genetic disease although its parents are only carriers of the condition. There is a one-in-four chance that every child born to two people with the same recessive gene will receive a double dose of the gene and be affected. This is how it works:

The chromosomes exist in similar, or homologous, pairs. Each trait is controlled by a double gene or a double gene sequence—one gene residing on one chromosome and the other on the other chromosome at the same place. When sperm and eggs are formed, the chromosome pairs split apart. One sperm or egg gets the genes on one chromosome; the other receives the genes from the other chromosome.

A recessive gene—the Tay-Sachs gene, for example—exists at a certain spot on a chromosome. Its partner at the same spot on the other chromosome of the pair is normal. Thus, in someone who carries the Tay-Sachs gene, half of the sperm or eggs have the normal gene, half have the Tay-Sachs gene. At the moment of fertilization, a sperm and egg come together and the newly created individual has a full complement of genetic material—half from the sperm and half from the egg. The genetic makeup of this new person can reflect several possibilities.

If two Tay-Sachs carriers mate, there is a 25 percent chance that a normal egg and a normal sperm will come together, creating an individual who has no Tay-Sachs gene and need never worry about his or her children being afflicted with the disease.

There is a 50 percent chance that an abnormal egg or sperm will fertilize a normal egg or sperm, creating an individual with one Tay-Sachs gene. This person won't be affected with the disease, but will carry the gene all his or her life. If such a person marries somebody without the Tay-Sachs gene, their children will not be born with the disease. But if the spouse also carries the gene, there is a risk.

Finally, when two Tay-Sachs carriers marry, there is a 25 percent chance that a sperm with the gene will fertilize an egg with the gene, giving the baby a double dose and causing the disease to be present. That's what happened to Gary and Marsha and their baby girl, Jennifer.

Most people are blithely unaware of their genetic heritage, of the possible deleterious genes they carry. Because of the nature of recessive disorders, a family can go for generations without knowing about the gene's presence. Gary's mother was the carrier of the Tay-Sachs gene in his family; Marsha's father had the gene in hers. Their aunts and uncles and cousins scattered all around the country had no idea that the gene lurked in their past. Jennifer was the first child to receive a double dose—at least in modern times.

In the daily course of their work, geneticists often see the result of people's ignorance of their genetic heritage. Geneticists recognized early that one means to deal with this might be to identify particular populations at risk for some genetic malady and attempt to screen them before they have children—to uncover the hidden deleterious genes and at least apprise a person of the risk so he or she wouldn't reproduce blindly.

Tay-Sachs is one of the genetic diseases that touches a specific population. People of Jewish descent whose forebears came from central and eastern Europe are at risk. They are easy to identify, and a simple blood test can reveal whether they carry the gene. Another such disease is sickle-cell trait. There are an estimated 2 million blacks who carry the sickle-cell anemia gene. A simple blood test can reveal its presence also. Although there are no tests yet to unmask the cystic fibrosis gene, an estimated 10 million whites carry it. When a simple blood test is developed, and there is optimism that it will be, a large segment of the population might then benefit from mass screening for cystic fibrosis.

Developments in understanding other genetic diseases that cause disorders in the body's chemistry also led to the realization that the effects of some genetic disease could be reduced or even eliminated if the presence of the disease could be spotted within hours or days of birth through blood tests. One of the first diseases to respond to this approach was phenylketonuria (PKU), a disease which causes mental retardation but which can be prevented through special diet in infancy and childhood. A simple blood test can pick it up.

In the 1960s several states began to require PKU testing. Soon other diseases yielded to this approach. In some states there are now compulsory newborn blood tests that look for a half dozen or more diseases, including sickle-cell anemia and hypothyroidism.

Nevertheless, the idea of screening various ethnic groups for a genetic defect, however laudable the intentions, brings chills to many civil libertarians. They worry that the results of the screening can be used for more than simply apprising an individual of the presence of a deleterious gene. Beyond such questions, there is also doubt about the efficacy of mass screenings—the cost, the ability to reach target populations, and availability of medical personnel to carry out the screening and then properly counsel those in whom an unwanted gene has been discovered.

Questions such as these led the Albert Einstein College of Medicine to embark on an experimental Tay-Sachs screening program in the early 1970s.

"There has been a lot of concern about the ethical questions raised by mass genetic screening," Nitowsky told me. "I certainly see some problems from my point of view. There is always difficulty getting the people who need the tests for a genetic condition to submit to them. Poor compliance leads to talk of mandatory tests. I personally object to this approach and so do many others.

"Screening for Tay-Sachs has not received as much criticism as some other kinds of screening programs, notably testing for carriers of sickle-cell trait among the black population. In that case I think there was inadequate education of the target population. People who were revealed to be carriers of the trait misinterpreted the information and thought they had the disease. A certain stigma came to be attached to being revealed as a carrier —again mistakenly. These people couldn't get insurance. They were refused jobs. Moreover, a problem with sickle cell is that once you have two carriers who are married, you can't offer them a lot. Amniocentesis won't reveal sickle-cell disease. There is a new experimental procedure called fetoscopy with which sickle-cell disease has been diagnosed in some fetuses. But it's not widely available.

"The geneticist must be very sensitive to his patient when it comes to discovering things about genetic makeup. We geneticists are attuned to biological differences, but we must pay attention to cultural and social differences, too. Suppose you give someone information about their risk of being a carrier of a genetic condition. You tell them that their chances of being a carrier are one in 25. Then if you give them a questionnaire to fill out and ask them what is their risk, they may say one in 1,000. No one expects to be a carrier. It can't happen to me. So when someone is identified as a carrier, it can be quite disturbing at first. These people need counseling and explanations of what being a carrier means so they do not feel stigmatized. This approach generally works in adults. But I object strongly to mass screening of children, high school students, for example. They simply aren't prepared to cope with such information.

"Our Tay-Sachs screening program has been quite successful, but even so I don't think mass screening is the way to deliver this type of health service to the community. It really should be

part of an ongoing effort where the family physician is the principal advocate of testing. Even the clergy might be involved in
this.

"There is a great need among physicians for postgraduate education about genetics. This has been borne out in surveys which
show that medical school graduates who finished their training
ten years or more ago and never had a genetics course have a
lack of information about genetic disorders. They simply don't
appreciate the importance of screening their patients to determine if they are carriers of various conditions. Education is the
answer in the end—education of the health care providers and
education of the public."

After learning that they are carriers for a recessive genetic
condition, and that their daughter had received a double dose of
the gene and would eventually die from it, Marsha and Gary felt
an intense anger burning within them, building up to the explod-
ing point. Gary in particular searched for someone to blame. He
turned first to the pediatrician, the man who had been his childhood doctor.

"I called our pediatrician about two weeks later, because I felt
I was close with him. I said, 'Did you know?' He told me he
thought he had seen it some months before. 'I didn't see any
reason to tell you any sooner until she displayed more evidence
of the disease.' He said he wanted to let us love her and enjoy
her. He probably thought we wouldn't believe him anyway.
Maybe we wouldn't have. I asked him when we should bring
Jenny in next. He told us to call the Tay-Sachs Association. He
said not to waste our money bringing her back to him. He certainly washed his hands of us.

"I had and still have a lot of resentment about what happened
to us, because I feel it could have been prevented. I was just so
angry that we weren't given the option of knowing. Somewhere
along the line someone should have told us. But no one wants to
take the responsibility for educating people about this.

"I've heard some doctors say that informing Jewish couples
about Tay-Sachs isn't their concern because it's a religious thing.
We had a very progressive rabbi then. Actually he was my father's rabbi. We drove over to his house the day we found out

and he was wonderful. Later, we had other meetings with him and at one meeting I asked him why a rabbi will marry a couple without first telling them about Tay-Sachs and the tests, and where to get them and what the risk is. He said there is enough pressure on a couple before getting married and he didn't feel that it was the proper thing to do. This was an attitude he said a lot of his fellow rabbis had. He said that it is a medical problem.

"We later became involved with another rabbi who is quite progressive. He started some seminars to discuss Tay-Sachs. Had doctors come as speakers, and so on. But the attendance was poor. People don't want to know. They just don't want to know.

"I have a lot of guilt about not knowing about this disease. I have a lot of resentment about not being told. You might ask whether my parents were aware of it. Why didn't they say something? I don't think they knew about Tay-Sachs. To tell you the truth, I have never sat down with them to talk about it. I don't want them to feel I have any resentment against them."

There is a dichotomy of emotions swirling about the parents of a Tay-Sachs child. On one level is the guilt about what they have done to their child, the remorse, and the desire to atone. But on another level is the desire to replace their defective child with a genetically whole child and a subconscious wish that the Tay-Sachs child die.

Within six months of Jennifer's diagnosis, Marsha became pregnant. At sixteen weeks' gestation she went to Einstein for an amniocentesis. Harold Nitowsky stood beside her and held her hand while the obstetrician performed the amniotic tap.

It takes three weeks for the fetus' cells in the amniotic fluid to grow out in tissue culture so that the tests for the presence of an enzyme called hex A can be performed. The enzyme levels reveal whether the fetus is normal, a carrier of Tay-Sachs, or affected with the disease. But the Einstein doctors also use a preliminary test that can be performed on the amniotic fluid the same day it is removed from the uterus. This test indicated that Marsha's baby would be normal. The formal results three weeks later bore that out. Marsha and Gary's son—the amniocentesis had also revealed his sex—would not carry the Tay-Sachs genes at all. He need never worry about the disease striking his children.

Marsha's pregnancy was uncomplicated. Jennifer had been delivered by caesarian section and Marsha was scheduled to have her second baby the same way, on November 7, 1976, thirteen months after learning that Jenny had Tay-Sachs. As her pregnancy drew to a close, Marsha's doctors moved the delivery date up to November 1 and then into the end of October. Ironically, Joshua was born on the same day that Jennifer had been born two years earlier. The baby was normal and blood tests revealed that the amniocentesis was correct—he carried no Tay-Sachs genes.

"For about two weeks after Joshua was born, Marsha and Gary just suspended their emotions," Sandy Silverman recalls. "They felt disloyal to Jennifer by loving this new baby.

"A Tay-Sachs couple must work out their feelings. They experience difficult emotional gymnastics. You know that on one level they expect their affected child to die. But another part of them wants the child to keep fighting. Marsha and Gary had these feelings. We talked about it.

"There is one word you can't say to a Tay-Sachs couple. The word is 'replacement.' Replacement of the Tay-Sachs child with a new, normal child. Marsha and Gary wanted a girl so badly and Josh was not a girl. They were unhappy about it but didn't understand why. Marsha once told me with a certain amount of satisfaction that her sister-in-law's baby was a boy. Marsha wanted a girl, and the reason was she wanted to replace Jennifer. But it's very difficult for a Tay-Sachs couple to admit to such feelings. They don't want to be disloyal to their ill child.

"Each couple must work things out for themselves. Each couple copes differently. Marsha and Gary went far beyond the call of duty. They kept Jennifer at home long after most couples would have given up. After an initial period of grief, Tay-Sachs parents begin to think of institutionalization because their child will soon reach a vegetative state. But not Marsha and Gary.

"They had a special nurse who lived with the child. When the nurse was gone they cared for her. One of them had to be with her all the time. If one went out, the other stayed home. They stopped doing things as a couple. They had Jenny on a suction machine because she could choke. They learned how to feed her through a nasogastric tube. They used to take her onto their bed at night and play with her long after there was any response. But they imagined response. They imagined that she still recognized them long after she stopped.

"We would see them periodically and mention the option of not keeping Jennifer at home, but Gary wouldn't hear of it. Finally, Marsha began to crack. She called one day and said she was near the breaking point. She was ready to institutionalize Jenny, but Gary said no. It was about time for them to bring the child in for an examination by Dr. Nitowsky. He examined her and told them she had to be institutionalized. Gary finally agreed. It was as if Dr. Nitowsky released him from part of his guilt.

"When Joshua was one year old, which was also Jenny's third birthday, Marsha told me something that reduced me to tears when I hung up the telephone. She told me how they had split the day so that they could spend part of it with Jenny at the hospital celebrating and the other part at home with Josh celebrating. At first you think what a lack of being in touch with reality that shows. But then you realize that each couple deals with it in their own way. What we know is that a couple must work it out, assuage their guilt, make their sacrifice, cope as far as they can. Marsha and Gary have coped as far as anyone can. Further: all the ritual they've gone through I think in the long run will free them when Jennifer dies. It will free them to love Josh even more than they do now, to be able to have another child and to be able to be all right."

"We had a special nurse who lived with us," Marsha told me. "She lived in Jenny's room and stayed with us from Sunday night to Saturday morning. So Gary and I took care of her Saturdays and Sundays. We would sleep downstairs with her. We did that for two years because I had a fear that she would choke and I wouldn't be in the room. You just couldn't leave her alone. We practically lived in her room. During the day we would bring Josh in and play with him on the floor.

"The only problem we ever really had was fevers, which we learned was usual for Tay-Sachs children. To be frank about it, she became a vegetable. She just lay there with her eyes open, no movement, no brain activity to speak of.

"As far as the daily routine, she was being fed by a nasogastric tube. The tube had to be inserted three times a week. We went to North Shore Hospital and learned how to insert it so we could tube-feed her. But Gary was much better at it than I was. He was better even than the doctors in the hospital. Three evenings a week we took it out and she would sleep overnight without it. She was getting phenobarb as a tranquilizer and we would take the tube out after we gave her the phenobarb. Then Gary would insert it the next morning.

"The morning routine would start with bathing. The nurse did it during the week and we did it the weekends. Toward the end, before she went into the hospital, Jenny was very heavy. Gary would hold her in the tub while I bathed her. Then she had to be fed through the tube."

"Inserting the tube wasn't so difficult after the first few times," Gary said. "You have to put it through the nose and it goes down the throat into the stomach. You have to check that it's in the stomach and not the lungs. You take a syringe and put it onto the tube and blow some air into the tube. With a stethoscope you listen for the gurgling sound of the air in the stomach. If you don't hear that, it's not in the stomach. So you put the end of the tube in water and see if there's air coming out, which would mean you have the tube in the lungs.

"But I never got it in the lungs, fortunately. In the beginning, she was still fairly conscious of her surroundings as we inserted the tube. So the first few times she would scream a little, more from fright than pain. After a while, you become used to the feel of inserting it and it became a routine.

"To feed her or give her medicines you hook a syringe to the tube and inject them through the tube into the stomach. Being able to feed her that way allowed us to keep her at home another nine months or so—another year, really, which is what we wanted."

"With a Tay-Sachs child, all you can do is make them comfortable," Marsha said. "They require little, if anything. Her care was never difficult at all, other than the constantly moving her so she wouldn't lie in one spot and get bedsores, and the feeding.

She had to be moved every hour. We had a suction machine in the house. She did have trouble with her saliva and choking. So we had to suction her if she needed it."

"The rough time was in the night," Gary said. "Later on, when she began to have problems with the NG tube, we slept in the room with her. We took her out of the crib and brought her into the bed we had in her room. That way we could hear her immediately if anything was wrong—if she began to choke. On weeknights we had the nurse in her room. But we had an intercom in our bedroom upstairs and we heard every sound in Jenny's room. Marsha was always very worried that she might not hear Jenny choking. Marsha learned to listen for any change in the sounds coming from Jenny's room and she would wake up instantly. The nights were hard.

"Once Josh was born it became very difficult, particularly for Marsha. I was still all wrapped up in Jenny. I tried hard to love the new baby, but frankly it was difficult at first. As far as I was concerned, Jenny was going to stay at home until I was told it wasn't right anymore."

"There came a point where I felt it wasn't a healthy family situation anymore," Marsha interrupted. "It had been over a year since we really did anything or were out of the house together. We never went out together on a weekend. We never left her with anybody. Any activity, such as family holidays, either Gary would go or I would go. One of us stayed home. Jenny had a very low resistance to infection, so we didn't have any children of our friends in the house for more than a year.

"The reactions of our friends is another whole story. It was difficult for them to be with us. We were down about what had happened to Jenny. We were jealous of their healthy kids.

"I remember seeing a child that was five months younger than Jenny doing things Jenny never did. Not that I wanted any harm to come to that child, but it was so hard to admit that here is this child doing things my daughter never did. I was jealous.

"So we only took her out if she had a doctor's visit or to go to Dr. Nitowsky at Einstein. It was hard to travel with her. She would have seizures if you touched her. Fast, five-second things. She was better off being left alone than moved.

"Finally, everything started bothering me more than Gary. He was wrapped up in Jenny, where I was really torn between the two of them. That's when I started looking into how and where to get her institutionalized. It wasn't fair to Josh. He was never getting out. He was cooped up at home with us—tied to Jenny's room, too. It wasn't a healthy situation for any of us.

"The time I felt guilty the most was when we decided to put her in the hospital. I knew we had done more than most parents would do, but I felt guilty. How can a mother do something like that? But she's better off now. The hospital has a unit devoted exclusively to Tay-Sachs children. They get excellent care. Not a lot of care is needed, of course."

A few months after Jennifer was hospitalized, Marsha became pregnant for the third time. Late in 1977 she went to Einstein for an amniocentesis. Unlike the first time, this amniotic tap was difficult. The obstetrician drew blood on the first attempt and had to keep trying. It was a long and anxiety-provoking session. Marsha left the treatment room badly shaken.

But the preliminary results the next day revealed that the child was unaffected. Marsha and Gary celebrated quietly. The third pregnancy was particularly important to them, because Marsha had been told she could tolerate no more than three caesarian sections. This child probably was their last chance at parenthood. Originally, they had wanted a large family. Now they had lowered their expectations and would be content with two normal children.

The couple were unprepared for the call Nitowsky put in to Marsha at work three weeks later. He wanted her and Gary to come to his office as soon as possible. He said he had some bad news. The amniocentesis results indicated the baby Marsha was carrying would be affected with Tay-Sachs. A member of Einstein's obstetrics faculty examined Marsha and suggested a method to abort the fetus vaginally, thus giving her a chance for another pregnancy. On December 5 Marsha successfully underwent the abortion, but it was a devastating experience. She entered into a period of depression and despair that rivaled the emotional extremes that had plagued Gary two years before when he learned his daughter would slowly deteriorate and die.

Sandy Silverman saw the young couple reach for each other again—Gary strong now, holding up his faltering wife.

When we began talking in Marsha and Gary's den, it was early afternoon. Josh had learned to walk and was into everything. First Marsha and then Gary would jump up and rescue him from some impending disaster as he tugged at a lamp cord or pulled at a bookcase. The affection they held for their son was obvious. They were the epitome of doting parents. But it was easy to imagine their conflicting feelings generated by this handsome boy and Jennifer, who at that moment lay in a hospital a few miles away, nearly four years old but unsmiling, unseeing, unfeeling, and unthinking. Were they resigned? Were they bound by their guilt and secretly resenting her? Or were they merely pulling away, anticipating her inevitable death? She had already lived longer than most Tay-Sachs children.

"She's a very strong girl. It's amazing," Marsha said, a curious note of pride in her voice. "Congestion, which interferes with their breathing and eventually leads to pneumonia, is the main problem that all the children in the ward have. But Jenny seems to be in very good shape compared to even the younger children who are constantly choking and congesting. She's able to spit up the saliva and phlegm."

"We visit her several times a week," Gary said. "I go at night. Marsha makes it whenever she can. And then we go together on the weekends."

"My time there is maybe ten minutes or fifteen minutes. There's nothing to do. I hold her and I cuddle her a little bit. I'll kiss her. I'll talk to her, but of course I'm really talking to myself because she doesn't comprehend anything I'm saying. Sometimes I'll try to lift her. She's uncomfortable when being lifted. She has these seizures. On the weekends we both bathe her and I do lift her up and hold her for just a couple of minutes then."

"Bathing her gives you something to do," Marsha added. "I enjoy it. Somehow I feel better after I bathe her and fix her hair. She just looks so . . . pretty. She looks so terrific. It's like we put our own little personal touch on her. But other than that, it's very hard because she doesn't respond at all."

At the end of October, Jenny turned four and Josh two. I

spoke to Gary and Marsha again just a few days afterward. I asked how they had marked the two birthdays.

"We went to the hospital," Gary said. "We usually go on Sunday morning and the birthday happened to be on Sunday this year. But there was no celebration. In our own way we said 'Happy birthday, Jenny' and brought cards from the family and friends who do still remember. You buy her a new nightgown, which is really all she can use."

"When I visit her, I'm usually fine and under control," Marsha interrupted. "The only time I have a problem is on her birthday. I usually spend a half hour or forty-five minutes on my visits. But on her birthday I cut it short. It's a difficult time. When I was in labor with Josh and realized that he was going to be born on the same date as Jenny, I wasn't too thrilled about it. But everyone said it was a good omen and there's a reason for it. In a way now I'm kind of glad. I come home from the hospital and I'm forced to be happy for Josh. It kind of balances out."

I mentioned Sandy Silverman's comment about how Tay-Sachs parents either come together or move apart when they learn the bad news, and her memory of Gary and Marsha physically reaching for one another.

"We have become incredibly close," Marsha said.

"It forces you to either talk about it or hate each other for it," Gary said.

They confided that Marsha was pregnant again. At the proper time they would go to Einstein for another amniocentesis. From the hesitant way they discussed it, I knew they were trying not to think about the 25 percent chance that the baby growing in Marsha's womb had received a Tay-Sachs gene from each of them.

CHRISTINE

Patients who don't take advantage of the care offered them are a continuing frustration in the practice of medicine. Some high-blood-pressure patients don't take their medicines, although faithfully following a doctor's instructions can completely control the disease. Heart patients fail to stop smoking. The obese don't lose weight. The patient with a fractured finger won't keep the splint in place.

Geneticists aren't immune to such professional frustrations. A parent who carries deleterious genes may deal with the problem by pretending it doesn't exist. A woman will refuse amniocentesis even though it could help prevent the birth of a retarded child. She can't face finding the truth. If an affected child is born, a couple may deny the child's defect. Others who learn of a genetic condition for which they are at risk may deny physicians permission to contact and warn other family members.

Some patients either don't understand the genetic problem explained to them or hear only what they want to hear and spend years laboring under misconceptions. Sometimes they simply are incapable of understanding; other times their form of coping is to misunderstand.

At the University of Washington in Seattle, one family with myotonic dystrophy—a disease that involves wasting of muscles and loss of coordination, formation of cataracts in the eyes, and mental retardation or a gradual deterioration of mental acuity—was seen in the medical center's genetics clinic for work-ups on different occasions two years apart. The family, to whom I'll give the name Davidson, was first seen in 1973 after the mother gave birth to a child who died of respiratory failure and severe myo-

tonic dystrophy at the age of three months. Two years later,
after ignoring the possibility that amniocentesis might have told
her whether the fetus was affected with the disease, Mrs. Davidson gave birth to another child. It, too, had myotonic dystrophy.
As a result, the family came to the university's genetics clinic a
second time.

During the second visit, while an oral medical history was
being taken from Mrs. Davidson's mother, one of the clinic's
geneticists stumbled upon a startling piece of information. It ultimately came to involve Christine and her unsuspecting parents.

"Ordinarily, a child with myotonic dystrophy shows no signs
of the disease at birth and it's only during adolescence or early
adulthood that you begin to see the symptoms," explains Judy
Hall, an associate professor of medicine and pediatrics at the
University of Washington. "But in perhaps 10 to 20 percent of
the cases where the mother has the disease the baby who in-
herits the condition will have what we call congenital myotonic
dystrophy. In that situation, the newborn often dies or else has
severe contractures, where the muscles are fixed and won't
stretch. There may be mental retardation and other problems.
There is a lot of speculation about what causes this form of the
disease at birth, but there is no definitive answer."

Judy Hall is a plumpish woman approaching middle age who
draws her long brown hair into a ponytail and ties it with bright
scarves. She is a merry person with an expressive face and an
animated manner of speaking that conveys how much fun she is
having as a physician. She has the kind of outgoing personality
that leads you immediately to wish she were caring for your children. She went to Wellesley and then medical school at the University of Washington. She studied genetics under Victor McKusick at Johns Hopkins.

"We first saw Mrs. Davidson in 1973 when the child that died
was born. There's a genetics clinic at the Children's Orthopaedic
Hospital and Medical Center, which is just down the road from
the University of Washington Medical Center. The Children's
Hospital clinic is where we generally see patients with congenital malformations. The genetics clinics at the university usually
handle adults.

"The Davidson baby was seen in the Children's Hospital first. We diagnosed the myotonic dystrophy and recognized that the baby's mother also had muscular problems, almost surely myotonic dystrophy. The whole family was subsequently seen at the university clinic and had an extensive work-up performed. We found that the disease could be traced back several generations into Mrs. Davidson's side of the family.

"We saw Mrs. Davidson's father and he clearly had the disease, although he denied having it at the time of examination. Although we didn't see them, Mrs. Davidson had a sister and a paternal aunt who were said to have the disease. In taking their medical history, we learned that Mrs. Davidson's great-grandmother and great-great-grandmother had cataracts of the eyes, which might mean they also had had myotonic dystrophy.

"In some myotonic dystrophy families it is possible to offer amniocentesis as a means to detect an affected fetus. The reason is that although we don't know anything about the gene that causes myotonic dystrophy, we do know that it is located on a chromosome adjacent to another gene called the secretor gene, which we can test for. This is a relationship geneticists call linkage. You may not know anything about how the gene you're interested in works. But because you know it's present on a chromosome right next to a gene for which you can test, you can make inferences about the presence of the gene in which you are interested. So linkage becomes a powerful tool we use in predicting the presence of an abnormal gene.

"The secretor gene determines whether you secrete certain substances present in your blood into other body fluids, such as the saliva or tears. These substances are called antigens and they are related to your blood type. If you have AB-type blood, for example, then the A and B antigens will be secreted into your other body fluids if you are secretor-positive. If you do not have the A or B antigens in your body fluids, then you are secretor-negative. You can test for the presence of these antigens in the saliva and determine whether an individual is a secretor or not. The picture is even more complicated because even though a person has the secretor gene, that doesn't mean that person also has the myotonic dystrophy gene located next to it. It's usual to have the secretor gene and be perfectly normal.

"When we have a patient with myotonic dystrophy, we test that person and his or her relatives to detect whether they are secretors. You need at least two generations to determine how the two genes are linked in a particular family. If you can show in a family with myotonic dystrophy that everyone carries the gene for being secretor-negative or secretor-positive, and that that type of secretor gene is always being passed along with the myotonic dystrophy gene, then you can begin to predict things about that family. The secretor tests revealed that in Mrs. Davidson's family the secretor-positive gene was linked to the myotonic dystrophy gene. That meant that in her subsequent pregnancies amniocentesis might reveal the status of the fetus, as to whether it had myotonic dystrophy. During pregnancy the fetus' blood-group antigens are secreted into the amniotic fluid if it is a secretor, so you test the fluid and determine whether the unborn baby is a secretor or not.

"We told all this to the family in considerable detail in 1973 and encouraged them to come in for further studies if Mrs. Davidson became pregnant. But two years later another baby was delivered and it also was brought to Children's Hospital with congenital myotonic dystrophy.

"It was hard to know what it meant to this woman that she was having these affected children. But it concerned the house staff at Children's Hospital and it concerned those of us in the genetics clinics as well as the doctors taking care of the family. You would think with this kind of information available that the family would take advantage of prenatal diagnosis. But the question was, to what extent did the family really understand what was available? Frankly, they were a little dim and maybe they didn't understand what they had been told. The other question was, did they understand it and then choose not to use it, which was their right, of course.

"So the upshot of it all was that we saw the entire family in genetics clinic a second time. It was while we were taking the medical histories that Mrs. Davidson's mother in a sort of 'Oh, by the way' manner mentioned that, seven years before, her daughter had given birth to a girl. It was before she was married, so she gave up the baby for adoption. She had two alliances at that time—one with her present husband and one with another man.

Either man could have been the father. The other man wasn't available for testing.

"You can imagine how much this concerned us. Somewhere out there was a seven-year-old girl living with her adoptive parents, everyone probably totally unaware that the girl had a 50 percent risk of developing myotonic dystrophy in her teenage or early adult years. What should we do? Ignore it? Or try to find her? It created quite an ethical dilemma for us."

All fifty states, the District of Columbia, Puerto Rico, and the U. S. Virgin Islands have laws requiring that adoption records be sealed after the legal formalities of adoption are completed. This is done to protect the privacy of those involved. There are an estimated 5 million adopted persons in the United States. Almost two thirds are the result of illegitimate births. Nearly half of all illegitimate births are to women between the ages of thirteen and nineteen. Secrecy avoids embarrassment for young mothers who made a mistake and want to resume life with a clean slate. It protects the child from possible embarrassment. It protects adoptive parents from future disruption of their relationship with the child should the natural parents change their minds and set out to find their child.

Only three legal jurisdictions—Washington State, Kansas, and the Virgin Islands—permit an adopted child to inspect his or her adoption records upon reaching legal age. Six other jurisdictions—Alabama, Connecticut, South Dakota, Puerto Rico, Oklahoma, and Virginia—permit opening of adoption records to the adopted child and the adoptive parents when the child reaches legal age. In all other areas, records can be opened only by court order.

Over the years, courts have frequently denied attempts to have adoption records opened. A New York court said no to a husband who wanted records opened to verify prior adultery on his wife's part. A defendant in a paternity suit who wanted records opened to show that the woman involved had previously claimed another man was the child's father also was denied. A California court denied the request by a New York–based charity to open adoption files to locate beneficiaries of a trust fund.

Cases where a lower court's denial has been overturned by an appellate court are rare.

In the state of Washington, the Seattle geneticists discovered three possible avenues through which they might alert the adoptive parents of Mrs. Davidson's illegitimate daughter. They were the family court in King County; the state of Washington's registrar, which also maintains copies of adoption records; or the adoption agency itself.

The family court would have to be convinced that good cause existed to open the records. The state's new public disclosure law seemed to be written so that adoption records weren't private, although the doctors suspected court action would probably be necessary to test the law and force the registrar to open the records. The adoption agency might be persuaded to contact the adoptive parents because of the agency's obligation to follow up on the welfare of a child it had placed.

After questioning Mrs. Davidson about the adoption of her baby girl, the doctors learned the identity of the adoption agency involved. It was a Seattle agency affiliated with a church. But before taking any action, the Seattle doctors agreed they first must convince themselves that they should seek the girl, rather than let the matter drop.

"We did a lot of soul-searching," Judy Hall recalls. "We talked about whether it was appropriate to contact the adoptive family and, if so, how to do it. We asked ourselves if there were rules we should set up to guide us. The answer to that was yes.

"We had a couple of conferences on the question among all the genetics clinic attending physicians and the fellows. It was decided that three senior geneticists should determine whether there was enough benefit to the adopting family to initiate an attempt to contact them. We agreed we had to think through very carefully how they should be contacted. Put yourself in their place. Seven years before, they had adopted this baby girl, and now, what a blow to have someone come along and say she might have a genetic disease.

"Right away, we could see some clear benefits associated with contacting them. If the child had the myotonic dystrophy gene, then there would be a time in her adolescence or early adulthood when she would begin to develop problems. But it would be possible with this disease that she could be subjected to a million-dollar diagnostic work-up trying to find out what's wrong.

"Secondly, with myotonic dystrophy you develop eye problems. If her parents were aware of this possibility, it would be something that they could watch for and get her early treatment. Myotonic dystrophy may cause a gradual dementia—a mental deterioration. So it would be important for the parents to realize what might be wrong if she began to experience learning problems.

"Finally, if she had myotonic dystrophy and it was of a later onset, she might reach adulthood and begin having children and be completely unaware of the risk of giving birth to affected babies. Knowing about her genetic problem, if she were affected, she could approach childbearing from a more informed standpoint. Of course, at that point, we didn't know if she was affected. We only knew she had a 50 percent chance of being affected.

"Those were rather obvious reasons in favor of contacting the adoptive parents. But as we considered the problem, we saw there were also some powerful negatives.

"By interfering, we would be breaking into an established family where they probably thought of their adopted child as a normal child. You would be saying, 'Look here. This may not be a normal child after all.' You have no way of knowing in advance how devastating this might be to the relationships within the family.

"There are some very strong privacy considerations involved. It is entirely possible a family will say, 'We don't care if this child has problems. We don't want to be bothered with it. And we don't want you to bother us either.' You have to be sensitive to the fact that the parent-child relationship may be different in an adoptive situation than with a biologic child.

"If the parents are the child's biological parents they are more likely to say, 'All right. I wish we didn't know about this, but we've got to be aware of it.' But with an adopted child they might say, 'Well, we never have liked this kid anyway. Adopting her was a mistake. Let's find a way to get rid of her.' That may be a feeling that's felt and not expressed, but it could have serious consequences. If the adoptive family were unstable for some reason, telling them their child is at risk for myotonic dystrophy could just push them over the brink.

"If the genetic condition involved had been a recessive one, where she would be a carrier but not affected, you could argue in the end that you shouldn't contact her. The chances of her marrying another carrier and then having an affected child seemed small. But since myotonic dystrophy is a dominant condition with a 50 percent chance that she would have it, and, if she did, a 50 percent chance that her children would inherit it, we felt that situation was different. So in the end, after much deliberation, we decided to attempt to contact the family. Gil Omenn was the one who investigated how to actually accomplish the contact, whether through the courts, the adoption agency, or some other approach."

Gil Omenn is tall, balding, and intense. He earned his medical degree from Harvard University and added a doctorate in genetics after coming to the University of Washington. He was a White House Fellow in 1973–74. After Jimmy Carter was elected President, he joined the President's science advisory staff.

"We decided to investigate all three methods of contacting the family—the courts, the adoption agency, and the state registrar," Omenn told me in his office in the Old Executive Office Building next door to the White House in Washington, D.C. "The registrar's office contended that the new public disclosure law in Washington State was not written to permit access to adoption records. We felt that to force the issue might open a Pandora's box, which we didn't want to do. So we concentrated on the adoption agency and the courts. We went to the Superior Court of King County, which has a family court division.

"The chief judge of the family court was a man named Robert Winsor. He and his wife both have an interest in health issues. I knew them through some political activities of my own within the Democratic party and through his campaign for the judgeship. At the time we started this I didn't know he was the chief judge in the family court, I just knew he was a judge. It was a happy coincidence to discover him in the family court.

"I went to call on Judge Winsor and he was very interested in our problem and basically supportive of what we wanted to do. A meeting was arranged at which we sat around a table in a conference room and spoke to all the judges of the family court division at once. I think there were about eight judges.

"First of all, they were dumbfounded that we might want to open adoption records. They just didn't understand the first thing about genetic counseling. They couldn't understand why we might want to open records for some genetic conditions, such as a dominant trait, and not open them for a recessive trait. We really had to conduct a sort of introductory lecture on medical genetics.

"We talked it all out—the need for us to be extremely sensitive in the manner we contact families, under what circumstances we might want to contact a family or not contact them. Finally, one or two of the judges wanted to understand this in great detail. They asked us to go back and think about how to really guarantee confidentiality of identities. How do you really estimate the costs and benefits of contacting a family? You don't know what the family's educational level is, what their previous medical experience has been. You don't even know if the child is aware that he or she is adopted. The upshot of it was that they asked us to prepare a sort of legal brief outlining it all.

"After that there were more meetings. We consulted the associate state attorney general, who is the university's chief legal officer, to make certain we weren't out of line in approaching the court in the first place. We had more meetings among ourselves to try to answer the judges' concerns. We prepared a document outlining the matter, and that was circulated among the judges.

"The procedure worked out required that each time we wanted to open an adopted child's records we make a detailed, explicit case about why we wanted to do it. We had to have something to offer the child and the adoptive family of sufficient benefit to override the possible risks of interfering. We were to be very sensitive in our contacts, to try and determine if the child knew he or she was adopted, to go through a family physician if possible, and so on. We set up a careful procedure to make certain we would never see the natural parents and the adoptive parents at the same time. We would keep their medical records separate and there would be no reference by name or number in the charts of one family to the other family. Only the senior geneticist handling the case would hold the detailed records of the case.

"Ultimately we reached an agreement with the Superior Court. But it took some time because such matters never seem to move quickly. Also, the judges didn't feel in any hurry because, as we knew and they pointed out to us, we could simply go ahead and contact the adoption agency and attempt to persuade the agency to help us make the contact. You could make a case for the agency doing that because of their obligation to follow up their placement of a child.

"Ironically, that's what happened. Long before we reached an agreement with the court, we had made contact with the family through the adoption agency. In fact, in a ten-month period we had three different cases like this and in each one we went through the agency. In each case, I might add, the agency insisted upon letting its personnel make the contact rather than us. In each case, we feel the way the news was initially given to the family wasn't handled well. Perhaps we should have spent more time with the adoption agency people handling the matter. But on the other hand, the agencies seem very protective of their personnel. Shall we say they don't like to take advice?

"Not long after this problem first presented itself to us, I was in Las Vegas, Nevada, to deliver a talk before a joint meeting of the American Medical Association and the American Bar Association about biomedical ethics. After my presentation, this man came up to me and introduced himself. He said he was a lawyer and that he also was a third-year medical student at the University of Washington back in Seattle. He asked if he could do an elective course with me bringing together law and medicine. I said, 'Yeah. You can help me with this adoption problem.' So we teamed up. He did much of the legal research and went to the meetings with me and did most of the writing of the brief. His name is Ken Hansen. In fact, I think it was Ken who ended up being the first person that Christine's adopted parents spoke to in the medical genetics program after the adoption agency's social worker broke the news."

"We have three adopted children. Christine is the youngest," says Christine's adoptive mother, to whom I'll give the name

Arlene Scott. "We told all our children as soon as they were old enough to understand that they were adopted. Charlie and I go through this whole story about how the thing that makes babies inside of me wouldn't work and had to be taken out. So we went to the agency and Daddy and I decided exactly what kind of little boy or little girl we wanted. We told them what we wanted and they looked and looked and finally one day they called us and said we have what you want. So we went down and looked, and sure enough, there you were—Sammy or Tony or Christine.

"When the kids were little, it was the kind of story they would ask to have told over and over. In a way, I think it's easier for a small child to grasp the idea that they came from a store than that they came from inside their mother. Going to the store and getting something, whether it's a loaf of bread or a baby, may be easier for a small child to relate to.

"As they got older they might ask, 'Well, if I didn't come out of you, who did I come out of?' We would tell them that we really didn't know who they came out of, but that we are religious and that God intended it to be this way and that's why we have you and we love you very much. So there's been no problem with accepting being adopted. They know they're loved and there is no question.

"When they are grown I know there is a chance they'll want to seek out their biological mother or father. I don't think it will bother me, be a reflection upon me, unless we are totally on the outs and the children and I aren't getting along. But I don't foresee that because we have a very close family."

I spoke with Charlie and Arlene Scott one evening near Christmastime in their large ranch-style home in a Seattle suburb that sprawls over a ridge commanding a view of Puget Sound. They, their children, and an older woman who is a relative had returned a few hours before from a long weekend in Portland, Oregon. The living room was decorated for Christmas with a tree, ornaments, and stockings hung on the mantel of the fireplace. The children and the older woman watched television in the living room, the children sprawled over pillows on the floor and wrapped in afghans.

Arlene and Charlie and I sat around the table in their kitchen-den and talked late into the night. Charlie has receding brown hair and a deep baritone voice, with a booming laugh. He is a man who needs only three or four hours of sleep a night, so he teaches school by day and holds down another full-time job at night, in addition to several entrepreneurial sidelines. He mostly listened to the conversation, occasionally interrupting with puzzling non sequiturs.

Arlene is a tall, heavyset woman who hid her brown hair under a kerchief. She speaks softly, but is an obvious take-charge type who is the dominant force in the Scott family.

"Sammy is the oldest and Tony is two years younger and then Christine is just a year younger than Tony," Charlie said. "I'll never forget the call we got from the lady at the adoption agency. She said, 'Well, I have a little girl here.' And I thought, 'Let's get to getting her.' We practically had two of them in diapers already. We were living up near the university then and I was going to school and Arlene was taking care of Sammy and Tony as well as two other little ones for another couple during the day. So we had four. I said, 'What's the difference if we add another one?'"

"We got Sammy and Tony when each was about two weeks old," Arlene continued. "But they didn't call us about Christine until she was six weeks old. The reason was that they thought she had a medical problem and wanted to check it out. When we went in, they told us that her mother had a muscular problem that basically had to do with her ability to grasp things.

"We spoke to the pediatrician that the adoption agency used. I can remember sitting in his office. He pulled out this book and read from it. There wasn't much to read. He said that what Christine had wasn't handicapping in any way and that if it affected her in any bothersome way it could be controlled with medicine so she wouldn't show any effects at all. Basically, he told us not to worry about it. Turned out, of course, he was wrong. He called her problem myotonia congenita. But she was a perfectly normal, healthy baby and we sort of forgot about it.

"Shortly after we got the first baby, Sammy, a new social worker came from the adoption agency to visit us. She was new on the job, and she's always been the one we dealt with since then. The first time she came out I was fixing coffee and holding Sammy in my arms at the same time. He was kind of fussy and she asked if she could help me. I said, 'Sure. Why don't you hold the baby?' And she said, 'Well, I'm not sure I know how to hold a baby.' I thought it was really funny at the time because here she was coming out to see how I'm doing and she doesn't know how to hold a baby. But she's very nice and we sent her Christmas cards.

"One day in 1975 this caseworker called me and she was sort of cryptic about it. She said that a medical problem had come up with Christine and the condition that we were told about when Christine was a baby had been misdiagnosed and that it was more complicated than we had been told. She said there were problems now and that the doctors at the University of Washington wanted to talk to us about it. I tried to ask her questions about it, but she wouldn't go into any detail. She was evasive. All she would say was that it was pretty complicated. She wanted us to come in and talk to her. But she was very busy, she said, and she couldn't see us, like, for about a week. So we made an appointment to go in and see her the following week.

"After I hung up I realized how worried I was. How upset I was becoming. I began to feel panicky. Here was this child that I loved so much and now perhaps something is going to happen to her. She's going to suffer or maybe not live a long life or whatever. I was panicky for Christine. I wanted to know what it was and if there was anything we could do about it. We would take her anywhere. Get her whatever treatment she needed.

"The adoption agency's caseworker had called early in the week and the appointment she made for us was the following Monday. By Thursday I was a nervous wreck. I couldn't stand it. I called her back and said, 'Look. I can't wait. I have to know something. I have to have a name for the disease. I have to know what it is. Can't you just give me a name for it? Can't I call the doctors direct?' She seemed different the second time I called her. Less reluctant. So she got out a letter or something the doctors had sent her and read it to me. She gave us a phone number to call.

"I called right away, of course, and spoke to Dr. Hansen. And he explained a little bit about it. He gave it a name. He told me about the 50 percent risk and that maybe Christine wouldn't have it or be only mildly affected. He said they wanted to do some tests on her to see if she had it or could pass it on to her children. They just wanted to talk to us about it so we'd know what was happening. So I said that of course we'd like to come in. I know I called them several times and had conversations with people there as I thought of more questions. They were all just very nice. They made appointments for us to come in right away because they understood my concern. They did all they could to make us feel comfortable. The only time I was terribly uncomfortable with it all was when the agency called and just laid it on us and then left us hanging. I told the doctors when I saw them I wasn't happy with that and I think I could tell that they weren't happy about it either.

"When I was talking with the university people on the telephone I told them I was greatly concerned that we not meet the biological family. We didn't want to know anything about them other than medical facts that might help Christine. When we went to University Hospital for the tests, we had to go to an office downstairs and be admitted. You go into a little cubicle and answer questions. There was a girl there sitting at a computer screen of some kind taking it down and I told her I was very sensitive about not finding out who the biological family was or the biological family finding out who we were. I told her there was to be no names, no contact. She said, 'Oh yes. Don't worry.' I don't even know if she understood what I was talking about. But she put in some information and then called up some information from the computer onto the screen, and I know if I had moved over just a little in my chair I could have seen the screen. Maybe nothing about the biological family would have been on the screen, but it really scared me. I absolutely didn't want to know anything about them."

It was mid-October of 1975 when Arlene Scott took Christine to the University of Washington Medical Center. She was examined by Judy Hall and Sandra Davenport, a physician participating in a postgraduate program in developmental biology.

Christine was given a thorough physical examination in which the physicians looked especially for any signs that her muscles were contracting and relaxing improperly. Special attention was paid to several reflex points, such as the one that causes the lower leg to jerk when the appropriate spot just below the kneecap is tapped. If myotonic dystrophy is present, the reflex point will respond to a tap, but the leg won't return to normal as quickly as it should. This test was carried out in several places, including the tongue. If the disease is present, tapping the tongue causes a depression that lingers for several seconds. The doctors asked Christine to perform common neurological tasks, such as swinging her arm in a wide arc and touching her index finger to her nose.

Then Christine was taken to the medical center's ophthalmology clinic, where her eyes were dilated and a lamp that projects a narrow beam of light was used to view the inside of each eye. The doctors were looking for any signs that the formation of cataracts might have begun.

Next she was given a thin sheet of a special paraffin to chew that would stimulate the flow of saliva but not introduce any outside chemicals. Christine chewed it dutifully, making a face at the feel of the wax in her mouth, and then spit and spit into test tubes until the doctors felt they had enough saliva for the laboratory studies.

Finally an appointment was made for her to return a few days later and undergo an electromyography test in which small needles would be inserted in leg muscles to measure electrical activity within the muscles. Abnormal activity might indicate the first deterioration of the muscles caused by the disease.

Writing in neat longhand in Christine's medical records folder after Mrs. Scott and her daughter had left, Sandra Davenport noted:

> A 7-year-old girl at risk for myotonic dystrophy with a completely normal physical and neuro-muscular examination.

Mrs. Scott was told she would be called to return for a conference with the genetics staff when all the laboratory results were available.

"I remember being a little bit irritated the day I saw Mrs. Scott and Christine," Judy Hall recalls. "I knew about the case and had been involved in the discussions, but I thought Gil Omenn was going to see them in clinic. So I wasn't fully prepared to see them that day. I wasn't as totally conversant with the case as I should have been. But things are never ideal in a clinical situation, and it worked out all right because we discovered that the Scotts are super people.

"While Sandra examined Christine and took care of her examination by the ophthalmologists, I sat down with Mrs. Scott and tried to explain some things and answer her questions. She made it clear that she didn't want Christine to know what was going on. I think it was an important insight on her part to want the child excluded from the discussions until something was confirmed or not confirmed. I explained all about the disease, that many people had myotonic dystrophy and lived useful lives, but that some people developed eye problems, and muscle problems might limit their activities.

"She told me what the pediatrician at the adoption agency had told them seven years earlier about the problem being myotonia congenita, which we already knew from the adoption agency's records. The Scotts had been told Christine had a benign disease, no problem, a full life. So it was clear that this doctor must have just spoken off the cuff and not looked it up or gotten into it. Both myotonia congenita and myotonic dystrophy are dominant conditions. So no matter what, Christine would have been at a 50 percent risk of having it. Seven years later, all you could say was that this man had been mistaken.

"Mrs. Scott was very nice, but I sensed that she was putting up a tough front, that what she was going through was very difficult for her. As we talked, she began to verbalize some of her feelings. She was unhappy about the way they were contacted and told about the possibility of a problem. Dealing with this obviously had caused some stress in her life and the family's life. I explained that we were new at this business of contacting adoptive parents, too, and that after she had thought about it we would appreciate any comments she had about how to handle it in the future. I explained how all the various tests we planned would work and how we probably could come up with a reasonable answer about whether Christine carried the myotonic dystrophy gene or not.

"Unfortunately, as often happens in medicine, when all the test results were in, the answer was rather equivocal.

"Christine's physical exam was entirely normal. She was a normal seven-year-old. In the slit-lamp eye exam, I thought I could see the slightest hints of something abnormal there. But it wasn't enough to be diagnostic. It was more my hunch that I could see something, but certainly not enough to advise the family upon. The electromyography was entirely normal, but that didn't mean abnormal electrical activity in the muscles might not be found as she grows older.

"Most disappointing of all, the secretor gene studies, which might tell us whether she had inherited the myotonic dystrophy gene, were also inconclusive. The difficulty was the identity of her father. If he was the man to whom her natural mother was married and whose secretor gene status we knew, then she hadn't inherited the gene. But if her father was the other man, whom we hadn't studied, then we couldn't say. Other blood-group typing didn't confirm who the father was, either. It was very frustrating.

"We had Mrs. Scott return for a formal genetic counseling session near the end of 1975. I was there, Gil Omenn was involved, and Ken Hansen and maybe one or two others of the genetics attending physicians or the fellows. Mrs. Scott was very disappointed that we couldn't be definite, but she took it well. We explained that we wanted to see Christine for an examination every two years to monitor her progress. We had a discussion about what Christine should be told, because children know something is going on, that something is wrong. It isn't enough to tell them it's just a checkup. Mrs. Scott had decided she wanted to tell Christine something and we discussed the various approaches she might take.

"In the time between the two meetings, Mrs. Scott obviously had spent some time thinking about the matter. She had some comments about how such cases should be handled. She said certain things should be thought through before the family is contacted. They were things we certainly agreed with—benefit to the child and family, contacting them in a sensitive way. She expressed concern that there might be adoptive couples who wouldn't be able to handle the stress of learning that their adopted child had a genetic problem, that their marriage might crumble or their relationship with the child might be affected. She felt very strongly that we should try to get as much information on the family as possible, some insight into their social situation, before hitting them with the news. She was very articulate about it all."

"There were many times through all this when I was in tears," Arlene remembers. "I was terribly disappointed that we wouldn't know one way or the other. I remember standing out in the hall after the doctors told me they didn't know whether she had the gene or not. I was crying then. It was hard to take. I think not knowing is the most difficult part of anything. It still pops up in my mind from time to time. Something will remind me. We'll be driving down the street and I'll suddenly realize we're passing the university and I'll think back to how long it is until we go in again. And when we do, will they find something? In one sense I'm anxious to go back again and try to find out. But at the same time I'm afraid of what they might find.

"Yes, there were a lot of tears then. I was very nervous and apprehensive and then very disappointed. And now as we talk about it I'm getting more and more tense thinking about it. There is that possibility there for Christine. There's the possible heartbreak if she can't have a normal child, if she had to terminate a pregnancy. That would be very, very hard. There could be some really hard times for her. Oh, that hurts so much to think about."

After Arlene, Charlie, and I had been talking for about thirty minutes, Arlene rose and put together a mix cake that cooked rapidly and could be eaten hot from the pan. She served the children in the living room and then we ate the rest at the table in the den and continued talking. Our conversation ranged widely over their past and Charlie's various projects and anecdotes about the children. More than once Arlene steered the conversation back to Christine and her physical activities—gymnastics lessons, baton twirling. She stressed how physically active her daughter is, her coordination and grace. At one point she called Christine into the den and presented her to me.

Christine is a beautiful child. She has short brown hair with bangs. She wore tiny button earrings. She has creamy white skin and a round, open face with straight teeth. Her mother urged her to recite for me all her activities and Christine dutifully ran through the list. At her mother's urging, Christine demonstrated

how to turn a cartwheel. It appeared to be as graceful a cartwheel as you might expect from any nine-year-old, even though
she nearly crashed into a wall.

It became clear as our evening drew to a close that Arlene,
probably in an unconscious manner, was compensating for the
uncertainty of Christine's future by urging her daughter to pursue physical activities, to deny through her physical achievements that anything could go wrong, to reassure her mother at
each new plateau of physical maturation that nothing was amiss.
I suspected that Arlene and perhaps Charlie had suspended their
emotions about their daughter's future, placed them on a long-
term hold. If Christine did have the myotonic dystrophy gene
and in a few years it began to express itself, I sensed that the
couple's delicate emotional house of cards would come tumbling
down.

But I had clearly sensed something else within the household:
an open, close-knit family in which each member is secure in his
or her knowledge of being loved and cared about. I had no
doubts that they would weather the bad times, if they were coming, and emerge whole people on the other side. How right the
doctors had been to worry about the effect their news might
have on an unstable family. Arlene had put her finger on a key
problem when she expressed that concern to the geneticists.

I met Judy Hall the next morning in the university hospital's
cafeteria for a cup of coffee. I told her about my impressions of
the Scotts and Christine's physical activities.

"I'm not surprised that something like that is occurring. But
it's probably healthy, if it doesn't become too extreme. That's
Mrs. Scott's way of coping with the situation," she said.

We talked about other things, and then she observed, "We
physicians must never forget that every patient can teach us
something. We learned a lot from the Scotts about how to handle this problem of the adopted child. My intuitive feeling is that
there are far more children like this than we realize. My emotional response is thank goodness it was the Scotts and not someone else that we had our initial experience with. We were fortunate that it happened first to an articulate, sharp, caring, and
feeling family. They helped us get a sense of the right way to do

this. That feeling of being thankful it was them is something I remember with a sharpness as if it were yesterday."

Two months later, Arlene brought Christine to the hospital to have all the tests repeated. It had been more than two years since the first examination. When they were finished, Judy Hall could see no change from the previous tests. But in the back of the eye what she believes to be the first signs of possible cataracts were still visible, although they weren't distinct enough to make a diagnosis. Judy Hall's hunch that Christine may carry the myotonic dystrophy gene and that it will eventually manifest itself remains unchanged.

GLORIA

It is a wet Monday afternoon in April at the Yale–New Haven Hospital in New Haven, Connecticut, a teaching hospital for the medical school at Yale University. In a small private room of the obstetrical wing on the fourth floor, a woman reclines on the electrically operated bed. The back of the bed has been raised so that she can read and watch the television that hangs from a bracket near the ceiling.

At first glance, the woman appears young, perhaps only twenty or twenty-one. But there is a hint she might be older—a crinkling about the eyes when she smiles and suggestions of deeper, future lines about the mouth. She is, in fact, twenty-eight, in the prime of her childbearing years, although she is childless. In another four or five years she will be past her child-bearing zenith, entering the twilight years of fecundity, sliding toward menopause and, ultimately, old age. She is more acutely aware of this inevitable senescence of her body than most women her age. It is the combination of being at Yale, the test she underwent that morning, and the years of coping with her condition, which doesn't physically affect her but has touched all those around her.

I'll call her Gloria. She is from Nashville, Tennessee, where she works in a large hospital as a medical technologist, performing the laboratory tests that are such an integral part of modern health care. She is the youngest of six children born to a family in nearby rural Smith County, Tennessee. Her southern drawl is charming but occasionally jarring when a sharp nasal edge creeps in. Gloria's father, now retired, was a tobacco farmer. Gloria's mother died when Gloria was sixteen.

She has done well for a Smith County farm girl—almost a decade as a medical technologist, a nice home in a suburban housing development in Nashville, a husband who is a policeman studying for a law degree. They have worked hard. To complete the picture, they desperately want a child. Indeed, a baby is growing in her womb even now, has been for sixteen weeks. It is the reason she has come to Yale.

Gloria casts aside the women's magazine she has been reading and turns on the television with the remote control that hangs from the bed. A young woman is angrily talking to an older woman, her mother, perhaps, who looks stricken at the young woman's words. There is no mistaking an afternoon soap opera. Gloria flips the television off and rises from the bed, crossing to the window.

Outside, the wind borne by a storm sweeping up the Atlantic Seaboard lashes at the bare tree limbs that await the kiss of spring. Raindrops tap on the window. She can see parking lots and other buildings in the medical center complex. Lights are on in most of the windows. Somewhere in one of the buildings a lab tech like herself, perhaps a woman of similar age and aspirations, works in a room like the one at the hospital in Nashville—tables lined with reagent bottles, pH meters, spectrophotometers, and the like. The lab tech is reading a chart, making the calculations, subtracting the difference between the base line and the sample lines. Gloria is familiar with the CPK analysis procedure. It runs quickly. They took the blood samples at 9 A.M. The blood must have been in the lab by ten. The analysis should have been done hours before. Why hasn't she been told yet?

She has thought so much about the moment when someone will arrive with the news that it no longer seems to matter. She feels detached, as if someone is going to tell her something that she must only relay to another. If the CPK values are low, then she will check out the next morning and go home. She knows she will be happy, reassured, but the waiting has so numbed her that she has trouble imagining how that will feel.

High CPK values will mean her baby has Duchenne's muscular dystrophy. Its muscles will waste away and death will come in the twenties. If the CPK levels are high, then early in the evening she will be wheeled into a treatment room and one of the doctors will insert a catheter through her abdomen into the uterus. A carefully calculated dosage of prostaglandin F_2 alpha will be injected, causing her to go into labor. Twelve hours or so after that, she will abort her sixteen-week-old fetus. It will be her second abortion, and she has decided it will be the end of her attempts to become a mother. But the waiting and the anxiety have dulled even the perception of this possibility, clouding her sense of how that will inevitably color the rest of her life. It is something she has just stopped thinking about.

She wonders who will bring the results. Inge? Dr. Hobbins? Dr. Mahoney? Probably Hobbins.

"She got in late Sunday night, after midnight, the poor girl," Inge Venus recalls. Inge is the clinical coordinator in the high-risk obstetrical unit at Yale. She handles most of the contact with the patients prior to their arrival, making arrangements, taking care of the dozens of details involved when someone travels half-way around the world or across most of a continent to seek the unique diagnostic procedures available at the Yale unit.

"She was supposed to be in by 9 P.M. or so, which would give her time to get settled and a good night's rest," Inge told me, studying Gloria's chart. For the doctors, the memories of patients often blur after weeks and months of procedures, but Inge can recall details of each patient's visit and the outcome of their tests. "A plane had an accident at the end of the runway or something like that, and Gloria's plane was delayed and she missed her connection in New York. I think they had to route her through Pittsburgh. Someone who comes here late at night, especially on a Sunday when there aren't as many people here, will be lonely. And her husband couldn't come with her. So I got up extra early Monday morning so I could get to her room by seven-thirty.

"When I first come into their rooms and meet them, I already feel like I know these patients because I've talked to them so much on the telephone. The first time I heard of Gloria was a few weeks before when her obstetrician in Nashville called. I took down the details and then asked him to have his patient call me. I try to confirm what the doctor has said by talking to the patient. I explain about our policies and what is involved. In

Gloria's case she was a carrier of Duchenne's muscular dystrophy, and this was not based on a high CPK value but instead was because of a muscle biopsy she had gone to the University of Iowa to have performed. That showed she is a carrier for Duchenne's, and she also has two sisters who are carriers.

"She is a lab technician and was familiar with the CPK test and had a good idea about the procedure. I remember she was so calm and objective about the whole thing. You didn't have to drag information out of her. Her mind was set. She knew what she wanted. I talked to her a couple of times the week before she arrived just to make certain everything was all set. So by the time I walked into her room that morning, I felt I knew her."

Inge Venus was born in Germany and came as a child to the United States with her mother and stepfather. She is blond, with a ruddy complexion, a dazzling smile, and just a trace of a German accent to give her voice a lulling softness. It is easy to imagine her gliding into Gloria's room, her smile easing some of the apprehension that had been building in Gloria's mind.

"Considering what she had gone through the night before and that she probably hadn't had much to eat, she was well composed," Inge recalled. "When the husband comes along we try to get him to sleep in the same room so they will feel more at ease. But he couldn't come and she was all alone. I made sure she had had something to eat—not a heavy meal, but enough so that she wouldn't feel faint. And then I began talking about the most important thing of all, the procedure.

"I wanted to know if the procedure had been explained to her. Did they just tell her it was going to be a needle insertion, or did they explain what we really are going to do? In her case, she was pretty much aware of everything. She knew it wasn't just a thin needle but a very thick needle. I explained how there is a scope inside it with a fiber-optic tube that permits us to look inside the uterus. It's very important to tell her that while the scope was being inserted she would feel this pressure for a second and it would make her want to gasp for air and try to breathe and for a moment she would not be able to do that.

"She asked why and I told her that the scope has a sharp edge on the bottom. To be able to go through the abdominal wall and into the uterus, the doctor has to give it a hard, sharp push. It's real quick, but hard. It takes the breath away. I feel that many times that aspect of it isn't explained and it catches a woman by surprise. They get a novocaine injection at the site of the insertion so there isn't pain. It's the sensation of the pressure and gasping for air that frightens them. Then, with the scope in, it's over. There shouldn't be any pain. If there is, I tell them to tell the doctor. We don't want to work on someone while they're uncomfortable.

"Then I told her how the scope has a needle beside it that the doctor can extend to prick a blood vessel in the placenta and take samples of the blood. And then the blood goes to the lab. And she understood all that. I told her that the doctor isn't always able to get any of the fetus' blood and that there could be no results. And she knew that, too. So then we talked some about her husband, who couldn't come because of work or school or something. I think he's a policeman in Nashville. We talked about her trip and the difficult time she had getting here. All in all, she was remarkably calm and composed and knew what she was doing. I liked her very much by the time I left her room."

Development of the fetoscope and fetoscopy might not have been pushed forward if two Yale doctors hadn't had a casual conversation one morning in 1972 while participating in the Pediatrics Department's "grand rounds," a periodic teaching exercise at which most of the department's doctors gather and discuss interesting cases and new developments in their specialty.

"Do you think we could ever sample the fetus' blood?" Jerry Mahoney asked, turning to John Hobbins, who was walking down the hall beside him.

Hobbins stopped and stared. It was a startling idea. "No," he said slowly, "I don't see how you could."

Serendipity often has a part in scientific and medical discoveries. It was at work that morning at Yale as Hobbins and Mahoney talked.

John Hobbins is a short, muscular man who moves quickly. He has a broad smile and a manner of looking you in the eye that puts you at ease. Jerry Mahoney is tall and lean, with an angular face and a Lincolnesque beard.

Mahoney was born in 1935 in Washington, D.C., a year before

Hobbins in New York City. Each went directly from high school
to college and then medical school before starting up the ladder
of academic medicine that led both men to Yale. Mahoney went
to Cornell and then medical school at the University of Pittsburgh. He was a pediatric resident at Johns Hopkins in Balti-
more, under Victor McKusick, and then returned to Pittsburgh
to become the chief resident at Children's Hospital. By 1966 it
was time for the obligatory two-year stint as a doctor in the
Army. He spent it as chief of pediatrics at the army hospital in
Fort McClellan, Alabama. He came to Yale in 1968 as a medical
genetics fellow and was appointed assistant professor two years
later. He became an associate professor in 1974.

Hobbins went to Hamilton College in Clinton, New York, then
came home to New York City to attend New York University's
College of Medicine. He interned for a year at Greenwich Hospital in Greenwich, Connecticut, and then went to Yale in 1964
as a resident in Yale's obstetrics and gynecology department.
He's been at Yale ever since, except for two years with the Air
Force at Otis Air Force Base on Cape Cod, Massachusetts. In
1976, the year he turned forty, Hobbins was named director of
obstetrics at Yale–New Haven Hospital and assistant chairman
of the obstetrics and gynecology department.

The road to success in academic medicine is a difficult path,
and the ability to conceive and execute original research projects
is essential to promotion. Mahoney and Hobbins had never collaborated on research before, but Mahoney's question about
sampling fetal blood stuck in John Hobbins' mind and reap-
peared to tease him several times during the rest of the day and
again that night at home.

By 1972, amniocentesis for prenatal diagnosis during the second trimester of pregnancy was shedding its reputation as an experimental procedure. More and more obstetricians were learning to insert a needle through the abdominal wall into the uterus
and extract the straw-colored amniotic fluid in which cells
sloughed off by the growing fetus float. A number of genetic diseases could be diagnosed by this method, and the number would
undoubtedly grow as new advances were made in looking into
the cell's interior to spot defects of the body mirrored there.

But the possibility of actually taking a sample of the fetus' blood was even more exciting. That could open up a whole new array of genetic conditions that amniocentesis couldn't touch. There is the group of blood disorders known as hemoglobinopathies—sickle-cell anemia, which affects blacks, and the thalassemias that strike people of Mediterranean origin—which perhaps could be diagnosed. Duchenne's muscular dystrophy might be diagnosed. There is a whole class of disorders involving the white blood cell that a sample of fetal blood might reveal. Sampling fetal blood might permit study of fetuses whose growth was retarded. Analysis of their blood constituents could, perhaps, reveal the cause of the retardation. Undoubtedly, future research by others into the chemistry of the blood would make possible the diagnosis of additional genetic diseases if a few drops of fetal blood could be obtained. The list of diseases that could be detected during pregnancy, permitting elective abortion, if desired, might dramatically increase.

Thus, the more Hobbins thought about Mahoney's question, the more excited he became. Mahoney had hit upon a significant idea for a research project. Careers rise and fall upon such things. The next morning, Hobbins sought out the geneticist.

"I think we can do it," he said, "but if we do, we're going to have to find a very small instrument with a scope."

Hobbins and Mahoney weren't the first to think about inserting an instrument into the uterus to view the fetus and perhaps sample blood and tissue. As early as the 1950s, one researcher had experimented with inserting a viewing device into the uterus through the vagina and cervix. These early attempts necessitated rupturing the amniotic sac, terminating the pregnancy as a result. Other workers learned to insert the scope through the cervix without rupturing the membrane. It was during this period that some of the first breathtaking color photographs showing the development of the fetus were published in national magazines, revealing the incredible beauty of the first months of each person's existence.

In the late 1960s, researchers began inserting the viewing and sampling instruments through the abdominal wall or surgically cutting down through the abdomen to the uterus. One research effort resulted in samples of fetal tissue and fetal blood from the umbilical cord. The potential for fetoscopy was predicted.

Hobbins and Mahoney weren't aware of some of the most recent research in this area as they talked that day. But the outline of their experimental approach seemed clear. The scope would be inserted through the abdomen, just like an amniocentesis needle. With a fiber-optic tube to introduce a bright beam of light into the uterus and the necessary lenses, they would go cruising through the placenta, so to speak, searching for a blood vessel carrying blood to or from the fetus. The placenta is the point where blood vessels from the mother and fetus come together across a membrane, exchanging nutrients and oxygen from the mother's blood and wastes from the fetus' blood. The blood vessels there are well developed.

But what would they do once they had found a fetal blood vessel?

Simple. Build a needle into the scope that could be operated from outside. Find the fetal blood vessel, extend the needle into the blood vessel, and draw a sample of the baby's blood. Then they would need some method to confirm on the spot that the blood they had just drawn was indeed fetal blood. That way, if they had obtained maternal blood, they could search for another vessel and take another sample while the patient was still on the treatment table with the scope inserted.

They decided that the answer to that problem might be to capitalize upon the difference in size between a mother's blood cells and those of her growing baby. The baby's are larger. They would need a machine that could take a blood sample and compare the size of the blood cells, calculating what percentage was fetal and what percentage maternal.

Hobbins searched for a scope. He eventually learned of a Japanese-built instrument used by orthopedic surgeons to view the inside of knee joints before surgery. He visited the company to inspect the scope and bought one on the spot. Then he and Mahoney began consultations with engineers, seeking help to design and build the cannula that would enclose the scope and contain the sample needle. The entire apparatus would then be inserted into the uterus through the abdomen. It had to be as thin as possible.

Eventually, an instrument was ready. Then came the first tests on women. This was the most delicate phase of their research.

Among other things, they had to prove that inserting the scope and sampling fetal blood would not alter the course of a pregnancy—by causing a miscarriage, for example. Because of the risk, they couldn't experiment on pregnant women planning to give birth. So they turned to the women coming to Yale for abortions. Many of these women were asked if they would submit to tests with the new instrument before the abortion. That way, if something went wrong, the outcome of the pregnancy would be the same as an abortion. The doctors would also have a chance to examine the aborted fetus to see whether insertion of the instrument a day or two earlier had damaged the fetus. Without this type of research, their project doubtless would have reached a dead end. The experiments began to give Hobbins and Mahoney confidence that intruding into the placenta wouldn't interfere with the fetus. They began cautiously to relate their fetal blood sampling to genetic disease.

They had initial success diagnosing fetuses affected with the hemoglobinopathies, then hemophilia. In early 1977 they diagnosed their first case of Duchenne's muscular dystrophy. Late that year they felt ready to accept more women at risk for Duchenne's, although they emphasized to each patient that the procedure and the interpretation of the data should still be considered experimental. An article about their work appeared in a publication circulated by the Muscular Dystrophy Association, and Duchenne's patients quickly began appearing in 1978.

Gloria was the eighth patient at risk for Duchenne's to undergo fetoscopy to determine if her fetus had high levels of the enzyme creatine phosphokinase (CPK). If it did, it was likely her child would enter adolescence and adulthood with its muscles slowly wasting away. She would watch her child be reduced to a helpless invalid and eventually die in its twenties, just as she had watched her two older brothers die.

"It started with my brothers even before I was born," Gloria told me the day she waited in her hospital room at Yale. She fidgeted nervously with the edge of the sheet on her bed. She wore a bright cotton robe over pajamas and her hair had been brushed until it shone.

"They first diagnosed it as polio back in the 1940s. I guess polio was all a small-town doctor in Smith County, Tennessee, could think of. That was my oldest brother, Floridine. Then somebody sent my parents to Vanderbilt University in Nashville. They did a muscle biopsy or something and said Floridine had muscular dystrophy. Then later they took in my brother Key and got the same diagnosis.

"There were six of us kids, three boys and three girls. I was the last, the baby. My two older brothers who had MD are dead now. Floridine lived to be twenty-one. Key died in 1973 at the age of twenty-nine, which for someone with MD is really sort of old age. My third brother, Donny, who is four years older than me, is fine. He didn't get the MD gene. He lives in Smith County and works in Nashville.

"Janice is my oldest sister and she has one daughter, Vi, who runs a high CPK. So that means that Janice is an MD carrier and she passed it on to Vi, who is an MD carrier. My sister Evelyn also is a carrier. She has three boys, and two of them have MD. Tracey is the oldest. He was born in 1966 and he has it. Tony is a year younger and he's normal. Then Timmy was born in 1971 and he's got MD. So if you start counting in my family, of us six kids only one—Donny—escaped. And out of the four children my two sisters have had, one is a carrier and two have MD. Those are pretty dismal statistics.

"MD shows up early even without medical tests if you know what to look for. First, you see no strength in a child, even as a baby. They learn to walk slowly and they have difficulty with normal things a baby can do, like taking your hands and pulling up or turning over. I think at first doctors didn't know too much about it or they didn't tell us all they knew. In the beginning they told us it was something that skipped every other generation. It hadn't shown up before in my immediate family. My mother has a brother who is fine. My grandparents didn't have it. But I grew up knowing there was an inherited disease in the family. Then later I learned that it's always there. The women are carriers and there is a 50 percent chance that they will pass it on to their children. If you pass it on to your daughter, she will be a carrier. If you pass it on to your son, he will have the disease. But it didn't really hit me until I got married.

"I became pregnant the first time in 1974. I knew about amniocentesis, but I had never thought of having it done. But after weeks of being pregnant and worrying I decided to go ahead. Now that I look back, I can see I didn't give enough thought to getting pregnant.

"So I went to a doctor in Nashville who did the amniocentesis. He sampled the amniotic fluid and the baby was a boy. That meant I could toss a coin. Heads it would be normal. Tails it would have MD. I guess I sort of panicked. I went right out and had an abortion. I had planned on getting pregnant again right away, but something held me back. I waited. And then after four years I decided that I was getting to the age that if I was ever going to have children it would have to be now.

"In the meantime I heard of a new test, a muscle biopsy they were doing that could tell whether you were a carrier or not. If you have high CPK values in your blood, then you know you're a carrier. But my CPK has always been borderline. There has always been a possibility that I wasn't a carrier, that I had my abortion needlessly. So I went to the University of Iowa at Iowa City, where a doctor could perform this test. They took a piece of muscle from my leg and did the tests. They told me I was a carrier.

"That really hit me hard, at first, because I had always sort of told myself that I wasn't a carrier, that it hadn't happened to me. But then I thought, now that I know, if I get pregnant I can always have another amniocentesis. So when I became pregnant again, I called the doctor who had done the first amniocentesis. He had moved away. I asked around the hospital where I work and learned of another doctor, well respected, who did them. I went to see him and told him I was a suspected MD carrier and wanted an amniocentesis done, and he said he didn't know much about MD. So I had to explain about high CPK levels and enzyme leakage from muscle cells and muscle degeneration. He agreed to do it.

"Then in the meantime, he had been reading a magazine and saw an article about Duchenne's MD. 'I never would have read it if I didn't have you for a patient,' he told me. It was about fetoscopy and Dr. Hobbins at Yale. My doctor said, 'I know of Dr. Hobbins. I'll be happy to get in touch with him and see what he can do for you.'

"He did the amniocentesis when I was twelve weeks pregnant. Then it takes three weeks for the cells in the amniotic fluid to grow out. So when I was fifteen weeks pregnant he called me and told me that I was carrying a boy again. So I told him he had better call the doctors at Yale. And the next thing I knew I was talking to Inge on the phone."

There are an estimated 150 X-linked genetic diseases like muscular dystrophy, conditions sometimes called sex-linked diseases. The most famous is hemophilia. The most famous hemophilia carrier was Queen Victoria of England.

Russian history has always held a special fascination for me, particularly the years before the Bolshevik Revolution and the tragic story of Czar Nicholas II and his family. While browsing through a genetics textbook, I came across a striking photograph taken in 1894 when more than thirty members of Queen Victoria's family posed for a portrait. The path of Victoria's errant gene and the destinies it would alter could be traced there.

Victoria's hemophilia gene is thought to have arisen spontaneously, through a mutation in either Victoria or one of her parents. Hemophilia had been unknown in Britain's royal family before that. Victoria passed the gene to three of her daughters—Victoria, Alice, and Beatrice—and one son, Leopold. The daughters were carriers, but unaffected with hemophilia. Leopold had the disease.

Alice and Beatrice married into other European royal families. Alice passed the hemophilia gene to her daughter Alexandra, who married Czar Nicholas of Russia. The couple had a son, Czarevitch Alexis, who was a hemophiliac. The Russian monk Rasputin came to gain inordinate power in the Russian court because he convinced the czarina that he possessed supernatural powers that could cure Alexis. Rasputin's effect on the Russian monarchy contributed to some extent to the court's decline and the success of the Bolsheviks, although certainly there were other, more profound factors at work.

In Queen Victoria, as well as in Gloria's mother, the X chromosome carried the defective gene. Victoria's husband, Prince Albert, and Gloria's father were normal. Their genetic makeup

had nothing to do with the appearance of hemophilia or Duchenne's muscular dystrophy.

There are 46 chromosomes in humans. They exist in pairs. The
genes on one chromosome of a pair are basically similar to the
genes on the other chromosome. In males and females there is a
particular chromosome pair that determines an individual's sex
and is called the sex chromosomes. In the female, each of the
pair is known as an X chromosome. In the male, there is an X
chromosome and a Y chromosome.

When eggs and sperm are formed, the chromosome pairs split
apart. So an egg contains only 23 chromosomes, a sperm only 23.
At the moment of fertilization, when egg and sperm merge and a
new person is created, there is a reunion of genetic material and
the full complement of 46 chromosomes is present again.

Gloria, like all women, has two X chromosomes. One contains
a normal gene; the other contains a gene for Duchenne's muscular dystrophy. During the formation of her eggs, one egg will
contain the normal X chromosome, the next an X chromosome
with the muscular dystrophy gene. Half of her eggs are normal,
half have the unwanted gene. In her husband's case, half of his
sperm would have an X chromosome, half a Y chromosome.

If the sperm has an X chromosome, the baby that begins to
grow has two X chromosomes and will be a girl. If the sperm has
a Y chromosome, the baby will have one X chromosome and one
Y chromosome and will be a boy.

How does muscular dystrophy fit into this?

At the time of fertilization there was a 50 percent chance that
Gloria's egg had the normal X chromosome untouched by the
muscular dystrophy gene. If this was the case, a sperm with an X
chromosome would produce a girl and a sperm with a Y chromosome would produce a boy. The child would be destined to be
normal. But if Gloria's egg contained the X chromosome with the
bad gene, there would be problems.

If the sperm had an X chromosome, the child would have two
X's and be a girl. But one of the X's would have the muscular
dystrophy gene. The effect of the normal X chromosome upon
the defective X would be such that the girl would not be
affected by the disease but could pass on defective X chromosomes to her children. If the sperm had a Y chromosome, then

the baby would be a boy. There would be only one X chromosome in his genetic material—the defective one. Since it would exist alone and not have the moderating effect of the other, normal X, the muscular dystrophy gene could express itself unfettered.

Thus, in Duchenne's muscular dystrophy, as in all X-linked genetic diseases, the women with the bad gene are carriers, the men are affected.

When the Nashville doctor took the amniotic fluid sample from Gloria's uterus, the cells were placed in culture dishes in the laboratory and caused to reproduce and grow. They were harvested after about three weeks and examined under a microscope. The lab technician could see that Gloria's baby had an X chromosome and a Y chromosome. That meant it was a boy.

But examining the chromosomes under a microscope couldn't reveal whether the fetus' X chromosome contained a normal gene or the unwanted muscular dystrophy gene. There was a 50 percent chance that it was a normal X chromosome and that a normal boy would be born. There was a 50 percent chance that it was a defective chromosome and a boy with Duchenne's muscular dystrophy would be born. To learn which, doctors would need a sample of the fetus' blood so they could test for the enzyme leaking from the fetus' muscle cells.

"This gal Gloria was unbelievable. I mean, she was just tough," John Hobbins said afterward, sitting in his crowded office across the hall from the room where he performs fetoscopies. He had his feet propped on his desk, revealing stylish low-topped boots. "She came all the way up here from Nashville by herself. She didn't move a muscle. She said, 'You stay in there as long as you need to and do whatever you have to do.'

"She got on the table about 9 A.M. and we did an ultrasound scan. We spread a water-soluble jelly on her abdomen. You can use any contact medium, but oil gets on my shoes and stains them. So we use this water-soluble gel. We turned out the lights and I began passing the ultrasound transducer across her belly. The gel is so the transducer will move easily across her skin. The transducer sends out a beam of sound waves, which are reflected back and converted to a picture on a video screen. We're getting very sophisticated now in our ability to visualize the fetus and

spot gross abnormalities, such as missing limbs and the like. But
we use it in fetoscopy to determine the position of the fetus and
the location of the placenta.

"If the woman asks, we show her the picture on the monitor
and point out the baby, its head, hands, and such. You have to
be careful about that. If a woman happens to be initially am-
bivalent about the procedure, showing them the ultrasound can
cause problems, because it points out the humanness of a fetus
that they may later elect to abort. If they ask to see it, then of
course we show them. But you kind of get a feel for who wants
to see and who doesn't. If we think they don't want to see it,
we'll use it, point out what we see to one another as the need
arises, and then go on to the next thing.

"Sometimes the fetus will be in the wrong position. So we ask
the woman to get up on all fours on the table and we push and
pull at her belly to get the baby to move, but it doesn't hurt the
baby. When we get the fetus into the correct position, then we
use the ultrasound to measure the distance that the fetoscope
will have to be inserted. The ultrasound machine will draw a
dotted line on the screen and each dot is a centimeter. So you
know how far to insert it and don't risk overshooting. It will also
show you the angle at which you must insert it. In Gloria's case,
the fetus was positioned correctly.

"Then we turned on the lights again and painted her belly
with Betadine, which is an antiseptic. I gave her an injection of a
local anesthetic and then made an incision in the skin where I
was going to insert the cannula. I told her first there would be
some burning from the local anesthetic. Then I told her as I inserted it she would feel a lot of pressure and there might be one
point where she would feel a pinch, but the pressure should not
be construed as pain. I told her not to move if at all possible. If
they move, it can really foul things up. But most of the time
there is no problem. If we get in cleanly, then in most cases during the first ten minutes the patient feels very little.

"Once we got in, Gloria presented us some problems. We had
to go around a corner to get to a fetal blood vessel. The placenta
was almost jackknifed, what we call a fundal placenta. It came
down and around a corner so that it was like working in a small

hole that was wedged and closing in on us, like going between two walls with the vessel at the end.

"We missed the first two vessels and obtained only maternal blood. So the procedure took a little longer than normal. But the third one was fetal blood. We drew the blood out into small vials. Altogether we got about two tenths of a milliliter. We removed the scope and put a bandage over the incision and sent her back up to her room. The whole thing took about thirty minutes."

Inventing the technology to reach into a woman's womb and sample a few drops of a growing baby's blood is a feat worthy of mention but has no intrinsic value unless the blood can give definitive answers about the baby's condition.

With Duchenne's muscular dystrophy, the Yale doctors faced several problems. In the disease, defective muscle cells leak the enzyme creatine phosphokinase into the blood. The muscle cells eventually die and are replaced by scar tissue. The body's muscles slowly waste away, leading ultimately to death, usually in the teens or twenties. High CPK levels in the blood are thought to reveal the presence of the disease, even before the muscle deterioration begins to appear.

But what are "high" CPK values? It is common, as in Gloria's case, for a carrier of the disease to have nearly normal CPK levels, while other carriers might have high levels. Thus, once a sample of fetal blood has been drawn and the CPK level tested, where is the dividing line between normal levels and abnormal levels? How high must the level be to advise the mother that her fetus probably has muscular dystrophy? How low must it be to tell her that her baby probably will be normal? How convincing is the evidence that the CPK levels really are a precise mirror of the disease's presence?

When doctors wrestle with such questions, they have two recurrent nightmares. One is that they will advise a woman that her child has muscular dystrophy, but tests of the subsequently aborted fetus' tissues will reveal that it was normal. That's called a false positive. The other is that they will advise that the fetus is normal, but, after birth, tests will show the baby has muscular dystrophy. That is known as a false negative.

Beyond the fundamental question about the meaning of data obtained from fetal blood looms another, even more basic: what assurance is there that the biology of muscle cells in an eighteen-week-old fetus with muscular dystrophy is similar to that in a child or an adult?

"They are both serious questions, not to be taken lightly," Mahoney told me, sitting at the desk in an office in the obstetrics wing borrowed for a few minutes from a colleague. Mahoney's office and laboratory are in a distant building. His laboratory assistant, Rona Mogil, had hurried off with Gloria's fetal blood samples packed in ice to begin the analysis. Mahoney remained to talk with me.

"By the time development of the fetoscope had reached the point that we were considering how to diagnose Duchenne's, some evidence had accumulated in the medical literature that high CPK levels in the fetus are related to muscular dystrophy. The studies involved relating CPK levels from aborted fetuses or premature babies with microscopic study of their muscle tissues.

"We had been collecting data from fetuses known not to be at risk for Duchenne's. In some cases they were from women who had come in for therapeutic abortions and were willing to undergo fetoscopy to help advance our research. In other cases they were from women who had come in for fetoscopy because of hemoglobinopathies. We would measure the CPK in the fetal blood and relate it to biopsy of fetal muscle tissue after they had an abortion. We've accumulated data on more than two dozen cases this way and find that the CPK level in a fetus that is not affected by Duchenne's never goes above about 150 international units per liter. But it is very risky based on this data alone to say that any fetus with CPK levels above 150 has muscular dystrophy. You just couldn't make an assessment without more data.

"Then in January of 1977 we had a woman come to us with her first pregnancy who had two brothers affected by Duchenne's muscular dystrophy. We told her that we could do a fetoscopy, but we could only give her advice based on our studies of fetuses that didn't have muscular dystrophy. An amniocentesis already had shown her fetus was male. So there was a 50 percent chance it would have the disease. She told us that whatever we told her based on the fetoscopy, she still planned to have an abortion. So we sampled fetal blood and then took samples of fetal muscle tissue after she aborted. We sent the muscle tissue to Betty Banker at Case Western Reserve University, who is one of the world's best pathologists. We stuck in a couple of samples of normal fetal tissue also and asked her if any showed muscular dystrophy. She picked out the one from the woman with the two brothers with muscular dystrophy. And the CPK level in that fetus' blood was 540 IU. So there we had our index case. Since then we've found that CPK levels in fetuses affected with muscular dystrophy are five to seven times higher than the normals.

"The other problem was in the laboratory procedures. When you take a sample of fetal blood you also have at least some maternal blood mixed in. You also get some of the amniotic fluid mixed. All three of these have CPK in them. So it's not enough just to measure the CPK in the blood you draw during fetoscopy. You have to subtract out the other sources of the enzyme so the final number you get is only fetal CPK. We measure the CPK in the mother's blood and CPK in the amniotic fluid and subtract them. Then we make other assumptions and calculations to arrive at a fetal CPK number.

"In the end, what I can tell Gloria is where her baby's CPK levels fall, in what range, and what our overall results seemed to indicate. I emphasize that our procedure is still experimental. Everyone agrees we need more cases in which the fetal CPK levels are high and muscle biopsies confirm the presence of the disease to be absolutely certain about our results. Most of all, we need a few more 'index' cases, women who undergo fetoscopy and then have an abortion anyway and in whose aborted fetus Duchenne's is confirmed."

While Jerry Mahoney described the difficulties inherent in giving Gloria meaningful information, Rona Mogil was hurrying across the medical center grounds with the vials of blood nestled in a bed of crushed ice in a Styrofoam bucket. Her destination was Mahoney's laboratory in the Laboratory for Clinical Investigations building, where she planned to run the creatine phosphokinase assay. The CPK test is a common one in most medical laboratories now, and technicians can use a commercially prepared kit of reagents, which simplifies and speeds the test.

Creatine phosphokinase is an enzyme. Enzymes must be present to promote chemical reactions in the body. Creatine phosphokinase promotes the conversion of adenosine triphosphate to adenosine diphosphate and creatine phosphate. This is a chemical reaction critical to the performance of muscle cells as they extract energy from the food we eat and then perform the work of muscle contraction to make the body do our bidding.

Rona had little knowledge of the woman whose blood she was working with other than seeing her on the treatment table. She had been in the room with Hobbins and Mahoney, helping operate the Coulter analyzer, the machine that measures the size of the red blood cells in each sample the doctors draw and indicates whether they have pierced a maternal blood vessel or a fetal blood vessel. She didn't know, for example, that Gloria was a medical technologist like herself, or that she planned to have an abortion if the CPK values of the fetus were high, or that she had already aborted one fetus because of the disease that stalked her family.

But after three years as a medical technologist—the past year with Mahoney's group—Rona did have an acute sense of the responsibilities that rested on her shoulders. Mahoney would certainly come in and examine the lines on the graph paper to see if his calculations agreed with hers before he went off to give the patient the results. But the young woman was alone with her task as she prepared the blood, placed it in the vials, operated the spectrophotometer, and attended to the countless other details involved in the test.

She was shy and nervous when I visited her in the laboratory and watched her analyzing some blood.

"How much do you worry about making a mistake?" I asked.

"You think about it all the time," she said. "It's part of your job."

"Do you check yourself all the time?"

"I try to. When you are dealing with somebody's future and you know a baby's life is involved, you don't want to give them misinformation. If you can't give them anything definite, then you should tell them."

She was fussing with graph paper as we talked. She showed

how she used a ruler to measure the distances on the paper. As I was about to leave, she turned to me.

"I do this the best I can. It's part of my job," she said earnestly.

Rona began by dividing the blood into several batches so she would have backup samples if the test had to be run more than once. Then she put it in a centrifuge, a device that spins the samples at a high speed. This causes the heavier blood cells to settle to the bottom, leaving a liquid on top. She poured off the liquid, which would contain any CPK present, and put it in special vials in which the test reaction would take place. She worked carefully in the laboratory, surrounded by other people working on other projects—all part of Mahoney's various research interests.

The test Rona performed would cause the formation of adenosine triphosphate from adenosine diphosphate and creatine phosphate. The CPK present in the blood samples would provide the catalyst required. The progress of the reaction would be monitored by directing a beam of light through a vial of the reacting chemicals and measuring how much of the light could pass through. These measurements would be plotted by an automatic device on graph paper. Results from the samples of blood taken from Gloria's placenta would be compared to standard solutions, and the difference between the curves measured and converted into CPK values.

The CPK assay takes almost three hours. It was nearly 1 P.M. before the first results were available. Rona took the charts from the machine and studied them, a ruler in hand to make the measurements on the squiggly lines to determine the final results. But her practiced eye quickly spotted something wrong. The lines representing the blood samples were almost flat, indicating virtually no CPK activity. That wasn't right, even for a normal fetus. What was wrong?

She glanced at her watch. The afternoon was wearing on. Clearly, she would have to repeat the tests. She decided also to include past blood samples, kept in storage, for which the CPK values were known. Results from these could be compared with the samples taken that morning. She turned to the reagents again. Now it would be four-thirty or five before she had an answer.

Gloria had reviewed the CPK assay before she came to Yale. She knew it shouldn't take more than three hours. The feeling that so often surfaced to lull her, the "it can't happen to me" feeling, was with her again. She had always thought she wasn't a carrier. But she had been wrong. She was sure the second pregnancy would be a girl. Her husband's family was famous for producing girls. But it had been another boy. This baby would have low CPK values. Or would it?

The reality of all the uncertainties that dogged her loomed out there in the distance somewhere, shrouded in shadows. There was a 50 percent chance. Is the glass half empty or half full? She would always be a person who saw a glass half full. But if it was half empty, she would have the abortion, whatever the emotional distress it caused. There would be no question about it. She wouldn't even have to consult her husband. They had decided that issue long before.

"My mother died when I was sixteen," Gloria said, rising to pace to the window again. The rain was hitting the window harder now. "I was the only child left at home and Key was still living then. My daddy and I cared for him. A lot of that fell to me. He lived until 1973. So I've always lived with MD. I have an up-close view of it. I've watched my sisters marry and have children and then have to live with this disease. Evelyn is the one with three boys and two of them have MD. She was able to work full time until recently. Now she stays home with them. The twelve-year-old has already gotten to the point where he can't walk. I think it's been a couple of years since he could. He's in a wheelchair now, or he sits on the carpet and watches television. He went to school for a while but now he doesn't. The younger one can still walk some. He's about seven or eight.

"There's no mental impairment with MD. One of my brothers was extremely intelligent. My brothers went to school but not all the way through high school . . . only through the sixth grade or so because then they couldn't walk.

"Evelyn lives in Smith County. For a while there was a teacher that came in, but not anymore. She stopped sending the boys to school because of the provisions they made for them. They were in with retarded children and were treated like they were retarded, and my sister didn't think it was worth the trouble. They mainly let them paint and things like that. They weren't learning very much. So Evelyn thought she could do just as well at home. They can count and tell time and things like that. They are intelligent children. My sister's husband drives a school bus and farms. They live on a farm.

"I haven't talked to my sisters too much about MD. I haven't told them I'm pregnant or that I was coming to Yale for this test. Nobody in my family has ever given me any advice about having children, except my father. He has said something like forget it. I didn't tell them about the first pregnancy or that I had an abortion. If the test results turn out favorably, then I guess I'll tell them. I'll tell my father. He won't understand too much about the test. He's a reasonable and intelligent man but not up on scientific things. My father has been through a lifetime of agony with MD.

"My husband's parents are the ones who will be excited about this if the test is okay. My husband's father is a retired barber. He's sixty-seven, and all his grandchildren are girls. And I'm carrying a boy. Ever since we've been married it's been, 'When are you going to have a child?' After five or six years they gave up. But I didn't tell them I was pregnant. I don't want to get their hopes up. They know the MD problem is there, but I really don't know how well they understand it.

"The idea of abortion has really bothered me. It bothered me a lot when I had the first one and it bothers me now to think in a little while I may have to have another. But the idea of MD bothers me even more. I know abortion is controversial. There are these people who say I shouldn't be able to have an abortion even if the baby will have MD. I saw an article once in which some lady with a Tay-Sachs baby said that the people who are against abortion for something like that are the people who have beautiful and perfect children about to start kindergarten. I couldn't agree with her more.

"It may be easier to accept this abortion, the necessity for it, but it won't be any easier to go through."

Gloria had put her finger on one of the most troubling aspects of the abortion debate. What right have others to zealously impose their idea of life and morality on this woman? What right have they to decree that she should give birth to a child doomed to die a slow death as a young adult? How can they make such judgments unless they have helped care for the victims of this disease and seen the anguish it brought her family?

However, if Rona Mogil's lab results showed high CPK values, within a few hours Gloria would receive the dose of prostaglandin. The drug would induce labor and within a few more hours she would give birth to her baby, eighteen weeks too early. It could never survive outside her womb. But only a few weeks more and it might survive. Certainly it might live for a few hours or even a few days if supported with the high technology available in an institution such as Yale. Where should the line be drawn? Who should draw it?

Geneticists familiar with Duchenne's muscular dystrophy know of patients who routinely request amniocenteses and abort all male fetuses. They keep trying until a female is conceived. That ensures that an affected child won't be born, but it means that half of all the aborted males were normal. Should these people be entitled to seek abortions for all male pregnancies, knowing that 50 percent of the fetuses are normal? How high must the risk be for an abortion to be proper?

In the late afternoon I had to leave Yale for another appointment. I made arrangements to telephone Inge Venus at five-thirty to find out the test results. My plan was to return that night and talk further with Gloria. After I left the hospital I began to wonder what it would be like to see her that night if the results showed she had to undergo the abortion. In a manner of speaking, I would be a party to the abortion. As we talked I had supported her view that others had no business condemning her decision to abort if they hadn't seen muscular dystrophy firsthand as she had. But I began to think about the baby growing in her. It was moving now. Its heart was beating. It was fully formed. It had features. Would I be aiding and abetting a murder as the anti-abortion hard-liners would certainly say? I wondered how I would feel talking to her that night, what we might talk about. I began to hope for the easy way out: low CPK values when Rona finished her analysis.

When I telephoned at five-thirty, Inge didn't know the test results.

"There has been some delay in getting them," she said. "But Dr. Mahoney is over there now and then he was going to the patient's room to talk to her."

I asked Inge to go to Gloria's room as soon as Mahoney had finished and tell her I wouldn't arrive until about 9 P.M. When I did return, at about nine-thirty, Gloria told me what had happened.

A few minutes after my call, Inge telephoned Rona again and finally learned the results. Then, about thirty minutes later, assuming Mahoney had already seen Gloria, Inge took the elevator to the fourth floor and entered the young woman's room, her face creased in a large grin.

"You must be very happy," she said jubilantly, before realizing how nervous Gloria was and how worried she looked.

"What do you mean?" Gloria said, puzzled.

"You mean you don't know?" Inge asked. The look on Gloria's face instantly told Inge Mahoney hadn't been there yet.

"Do you have the results?" Gloria said eagerly, sitting up. "Have they finished?"

Inge knew Mahoney wanted to tell Gloria about the test results himself because of the experimental nature of the muscular dystrophy test.

"Hasn't Dr. Mahoney been here yet?" she asked.

"No. Do you know the results?"

"I can't tell you the results, but I know it's good news," she blurted.

Mahoney appeared in the doorway on her heels, smiling. "I guess you already know the results," he said. He offered his congratulations—and cautioned again that the results were experimental.

Gloria checked out of the hospital the next morning and flew home to Nashville. All indications pointed to a normal pregnancy. She would deliver a boy in the early fall. Seemingly, the doctors had successfully treated another patient. And there the story would have ended, except for something that happened in San Francisco.

Several medical researchers around the country played a part in the development of the fetoscope at Yale and the laboratory procedures to find out meaningful information from the fetal blood samples. One of them was Dr. Mitchell S. Golbus, an obstetrician-geneticist at the University of California at San Francisco. Mickey Golbus was using the fetoscope with his patients and conducting research in collaboration with the Yale doctors. Patients at risk for giving birth to Duchenne's babies were being studied, as were women at risk for other genetic diseases.

At about the same time that Gloria and her husband were beginning their career as new parents, Golbus notified his colleagues in New Haven of some disturbing results from his research. Two women carrying fetuses with low CPK levels had given birth to babies affected with the disease.

The news dismayed the Yale researchers. It meant that CPK levels couldn't be used to diagnose a fetus with Duchenne's. It meant their research conclusions were wrong. They had no choice but to back down from their earlier enthusiasm for the test. They could continue their research, but they would have to caution every patient that no conclusions could be drawn from their baby's CPK blood levels. And they sat down to the unpleasant task of drafting a scientific report for a major medical journal noting their failure.

"It looked so promising," John Hobbins said sadly. "We knew that with only one index case we were taking a risk. We knew we could be criticized by some of our colleagues, who felt we should have more index cases. But the need for a test for this disease is so desperate. . . ."

In medical research, of course, there are thousands and thousands of experiments that never work. A researcher has what seems like a good idea. He or she proceeds, perhaps investing years of effort, only to meet a dead end. It's only natural that many of the failures are never reported. But when research work is clearly promising, there is a push from many directions for the scientist to report the results. For one thing, it permits other scientists to attempt to duplicate the findings. If others also are successful, then the new research findings come to be accepted as fact. If others fail, then the research falls by the wayside.

The ethical approach in such matters, of course, is for the scientist to admit failure as soon as it is known, particularly where it involves a medical procedure that could alter the destinies of thousands of patients. For Hobbins and Mahoney to continue attempting diagnosis of Duchenne's patients would have been unethical. They moved at once to act on the San Francisco findings, although only two cases were involved.

"There's a continuing argument among medical researchers about whether to share data from your research with the patients involved," Mahoney told me after learning of the two false negatives in California. "I've always wanted to share data with patients. There are others who say you should keep the results entirely secret from patients. If they want to abort their fetus, then fine. If they want to carry it to term, then that's fine also. They can make their decision based on other information, not the research information, which may not be reliable. Well, we urge our patients to do that. But we share the data with them. They're always so hopeful that, even though it's research, you can still tell them something. So I like to share the data. But I've always made efforts to get the message across that the results represent experimental research. Of course, the emotions of this situation are such that the patient would like to grab on to things and hope along with us.

"The question now is where do we go from here? There are other possibilities for diagnosis, but much more research must be completed. For example, there are abnormalities in the white blood cells and the red blood cells of boys with Duchenne's muscular dystrophy that you don't see in unaffected children. So we're examining the possibility of looking for that in fetal cells. But the overriding question that must be answered is whether these differences that we see in cells are the primary effect of the Duchenne's muscular dystrophy gene at work or whether they are secondary effects—things we see that aren't directly related to the gene and the manner in which it expresses itself in the body. If it turns out that these differences in cells do indeed reflect a primary effect, then we may see progress rather quickly. If we're seeing secondary effects, then it could be a long way off.

"Our future Duchenne's patients will be briefed so they clearly understand that we're looking for things that will be diagnostic—that we can't give any answers. Now that we have these false negatives, I don't know how many of these women will want to take the risk with our results and continue their pregnancy. I suspect some will. What we will hope for will be women at risk for Duchenne's who will still want to participate in our research even though we can't really give them a definitive answer about their baby.

"It is not an uplifting experience to be wrong. On the other hand, it's not something about which John and I feel ashamed. It is research. Sometimes it works and sometimes it doesn't. One keeps struggling when it doesn't work. We're going to keep struggling."

I first learned that Hobbins and Mahoney's Duchenne's results weren't holding up from one of their colleagues at Yale. It was late in the evening, and an attempt to reach Hobbins by telephone for details wasn't successful. I spent an uneasy evening thinking of Gloria. Had she had her baby yet? Was it normal? Maybe she was one of the patients whose low fetal CPK values were meaningless. Had she spent the last five months of her pregnancy convinced that her son would be normal? How had the truth been broken to her? I had her telephone number, but I couldn't bring myself to call.

I thought of her sisters whose lives had been forever changed by their affected children. I thought of Gloria's determination not to follow in their steps, her courage to come to Yale alone to undergo an anxiety-provoking experimental procedure. I remembered her jubilant face in the hospital room when I saw her several hours after learning that the fetal CPK levels were low. Had fate played her a cruel joke?

The next day, John Hobbins returned my call. "No," he said, "Gloria was not one of the false negatives."

"Thank goodness!" I exclaimed. Although I had spent only a few hours with her, I realized how much I cared about what had happened to Gloria. Later, I felt a twinge of guilt about momentarily forgetting about those two false negatives, whoever they were. Their lives would be forever marked.

THE O'CASEY SISTERS

The beginning of diabetes in a child is sudden and dramatic. It is something parents never forget. Mary O'Casey remembers it well.

"It was at the very end of June in 1972," she recalls. "Frank had vacation and we were going to the Adirondack Mountains in upstate New York. We were fortunate in a way that we ended up not going, because in the mountains we wouldn't have had access to the doctors that we have here in Manhattan. Being on vacation and all, we might not have known what was going on."

We are sitting in the O'Casey family's living room in a small, crowded apartment at the northernmost tip of Manhattan, a neighborhood adjacent to a wooded park that sprawls over rocky granite outcrops and is a reminder of how Manhattan Island must have looked before the Europeans arrived. Climbing the hill from the subway, past the parochial school where the O'Casey children are educated and the Catholic church where they worship, I caught glimpses of the Harlem River, which separates Manhattan from the Bronx. High-rise apartment buildings in the Bronx's fashionable Riverdale section loom across the river, and the toll bridge joining the two boroughs at the point where the narrow Harlem merges with the mighty Hudson is visible through the trees.

It is a neighborhood that once was all Irish, and proudly so. Now the Dominicans and Colombians and other Hispanics are moving into the buildings. Many of the Irish are fleeing. The others watch apprehensively. The apartment buildings are lines of demarcation, with the Irish buildings striving to remain Irish. There is a feeling that if the line is broken in a building, then the whole building will go.

A member of Frank O'Casey's family has lived in the building where we sit since 1943. Frank and Mary's apartment is small and some of the walls need paint, but it is spotless. The furniture is old and hidden by slipcovers. A room that opens onto the living room through double doors and perhaps was meant to be a dining room serves as a bedroom. Two of the girls sleep there.

It is a hot afternoon in July. In a few minutes Frank will catch the A train in the nearby subway station and travel to his job as a computer operator with a brokerage firm in the Wall Street area. Frank and Mary are side by side on the couch, facing me. The children, Cynthia, Margaret, and Catherine, are in the room, too. They are beautiful girls with light hair and clear white skin dusted with freckles. Cynthia is eighteen, Margaret fourteen, and Catherine ten. They are friendly. Everyone is friendly. But there is tension in the room. I connect the tape recorder and Frank and Mary begin to respond to my questions. The tension remains.

It is obvious they are a close family. Their world is circumscribed by the ethnicity of their neighborhood, the church and school on the hill, and the effects of diabetes. I am an outsider come to probe their tragedy. Without realizing it, they have rallied together to face this threat, just as they probably have rallied each time diabetes struck. As Mrs. O'Casey speaks I know there is more, perhaps layer upon layer underneath, that she wants to say but can't. Perhaps she is even unaware of a desire to speak those other thoughts. I sense the same thing about her husband. They can't mention their secret thoughts and feelings in the presence of their children. I wonder if they speak about them when they are alone together. This family, like others I have seen, has dealt with their genetic heritage—and their altered destinies—on the most urgent level involving the mechanics of daily coping. But have they gone beyond that?

I push these thoughts aside and focus on what Mary is saying.

"We noticed that Catherine was going to the bathroom quite a bit. She was terribly thirsty. And she was tired and listless, which wasn't her nature. So we figured we'd better have her looked at. Catherine was only three then. We took her to the pediatrician and he took a urine sample. He came out and told me it didn't look good, that it might be diabetes. He wanted us to go for a blood test the next day.

"There had never been any diabetes in my family that I knew of. Maybe somewhere in the back of my head I had read that these were the symptoms. But I was shocked. And then he said we would have to wait for the blood test results to find out for sure. What a terrible, awful wait that was. Diabetes was foreign to me. I didn't know what to expect.

"The next day I took her down to a testing place—a lab. They took the blood and then I had to wait another day. He called me on the third day and said to get her right into the hospital. Her blood sugar was over 400. It was so out of control that she was in the hospital three weeks while they brought it back to normal.

"Looking back now, in light of all that I know, I can't understand why that doctor took two days diagnosing her. She should have been put in the hospital right away. Dr. Nicholson certainly wouldn't have let that happen.

"We decided to change doctors, which is how we found Dr. Nicholson, after Catherine went into shock at home not too long after she was diagnosed. She had a very severe viral illness and was vomiting and not eating enough. She was getting weak, but a small child doesn't know how to express that. And we didn't realize she was getting so weak. She was sitting in that recliner chair in the corner. All of a sudden she fell asleep and I tried to wake her and give her some orange juice with sugar. I couldn't get enough into her. Then she started fighting. Hitting with her hands. She had energy that was unbelievable. I still can't imagine her being so weak with low blood sugar and still having that much energy. She was almost violent. She wasn't all there mentally.

"I couldn't get the doctor who was treating her then on the telephone, so we packed her off to the emergency room. There was a foreign doctor in the emergency room, a woman. I think she was Indian or Middle Eastern. She had trouble with English. This doctor kept saying, 'Insulin. She needs insulin.' And I would say, 'No. No. Glucose. She needs glucose.' You see, her problem was that she had low blood sugar. There was too much insulin in her already.

"They finally got hold of the pediatrician and he told them, 'You listen to the parents. They know what they're talking about.' The pediatrician was a nice man, but he didn't specialize in diabetes. After that, we said, 'This can't happen again.' We asked around. A medical secretary we knew had heard of Dr. Nicholson. We knew of another Irish family in the neighborhood with two diabetic girls. They were using Dr. Nicholson and loved him.

"You know what Dr. Nicholson told us to do if we ever had an emergency again? He said to call the local police precinct and use his name. He said the officers will come around in a squad car and take her to the emergency room. Then we wouldn't have the problem of parking a car and all. He gave us his home telephone number. He told us to call the police and then call him. He lives just around the corner from the hospital. He said that by the time we got there he would have everything ready.

"Thank God we haven't had to do that. But it's there if we need it."

John Nicholson is a lanky, pipe-smoking Oklahoman with piercing blue eyes and a soft, slow voice. His specialty is treating children with diabetes. As we talk, he sips coffee from a "Star Trek" mug. Until he was promoted and had to move to another floor in the Babies' Hospital building of the Columbia-Presbyterian Medical Center in upper Manhattan, Nicholson had an office in which an entire wall was devoted to an elaborate mural painted by a junior high school class. He commissioned the class to paint it, and the result was an often puzzling but always delightful mix of unicorns and trolls and all manner of fantastic animals cavorting through a forest glade.

"It was painted by an eighth-grade class," Nicholson explained. "A friend of mine is an art teacher there. I asked the class for their concept of a hospital. I didn't get that at all."

Diabetes is a disease in which the pancreas, an organ behind the stomach, ceases to manufacture enough insulin, a hormone that aids in the metabolism of carbohydrates. Without enough insulin, the amount of glucose, or sugar, into which the digestive process breaks up the carbohydrates we eat begins to build up dangerously. The ultimate result can be coma and death if there is no treatment.

Doctors now recognize several different forms of diabetes. The most striking is that which affects children. The appearance of juvenile-onset diabetes is quick and vicious. Something attacks and destroys the beta cells in the pancreas that manufacture insulin. Steadily accumulating research data is revealing that certain children are genetically predisposed to a factor in their environment that leads to destruction of the beta cells and an insulin shortage.

Besides treating juvenile-onset diabetes patients, Nicholson also works at unraveling the mysteries of the genetics of diabetes. He supplies the patients, and his colleagues at Columbia and the New York Blood Center investigate substances in the blood called histocompatibility antigens. These substances, known commonly as HLA antigens, are the subject of intense investigation because they apparently serve as "markers" for genes that predispose to several diseases. In these diseases, geneticists know little about the genes themselves, but by testing for the presence of HLA antigens and studying how they are inherited, scientists can learn more about the diseases. In juvenile diabetes, the HLA antigen known as DW3 is the key.

Because they became patients of Nicholson's, the O'Casey family found itself involved in some pioneering HLA antigen research. They ultimately became research celebrities of sorts because the doctors found that the family's HLA antigens reveal a rare and exciting relationship between family members that promises to solve the riddle of why children get diabetes.

"Juvenile diabetes puts a huge burden on the parents," Nicholson says. "They're always sitting on a powder keg. Their child has a basic instability of metabolic regulatory mechanisms. The whole situation can go to pot very quickly. A child can become very ill before their eyes. So that's a big emotional burden.

"In addition, the parent has the knowledge that diabetes is a long-term disease. Insulin injections and watching the diet and exercise will control it while the child is young. But the parent knows that as the child enters adulthood the complications are going to set in. Their kids can look forward to a shorter life-span. They have blood vessel problems, eye problems, impotence. It's a serious disease.

"Thirty years ago diabetic kids were isolated. They weren't allowed to do things that other kids do. Now there is a much greater awareness among schools and camps and other institutions that diabetic children can do virtually anything that anyone else does. But I think many diabetic children still feel different, handicapped, tainted. I encourage my patients to do everything. So far I haven't had any problems with that. They go on wilderness trips and perform violent sports.

"I grew up in Oklahoma City. I went to medical school at Vanderbilt and did my internship there. I came to Columbia for my pediatric residency and then I was a fellow here in biochemistry. Most of my research was in nitrogen metabolism. That's not too far removed from carbohydrate and fat metabolism and diabetes. In 1965 I didn't have a clinical commitment, so when a diabetologist retired I took over his clinical involvement.

"Several years ago others here at Columbia began looking into the HLA antigens. There was growing evidence that linked the presence of various HLA antigens to some diseases. They were interested in studying HLA antigens in a group of people with the same disease. So I said, 'Why not juvenile diabetes?' I had a large group of patients who would cooperate in the research. The idea was to look at those families where more than one child had diabetes and search for a relationship between juvenile-onset diabetes and HLA antigens. Later, we expanded it to include families with only one diabetic child. That's when the O'Caseys became involved, because of Catherine. Dr. Nicole Suciu-Foca here at Columbia and Dr. Pablo Rubenstein at the New York Blood Center analyzed the blood for the HLA antigens."

Heart and kidney transplants were what led, really, to the involvement of the O'Casey family in the Columbia-Presbyterian doctors' research. Before transplanting an organ such as a kidney, surgeons look for genetic similarities between the donor and recipient. Similarity lessens the possibility of rejection. The donor-recipient compatibility is determined by histocompatibility antigens. These antigens are biochemicals expressed on the membranes of most cells, including those circulating in the blood. Their presence can be detected by various laboratory tests.

If a kidney is transplanted that isn't a good match for the person receiving it, the histocompatibility antigens expressed on the membranes of all the kidney cells are just different enough from those expressed on the cells of the recipient to cause the recipient's immune system to mount an attack and reject the kidney. Attempts to refine detection of these antigens and measures to obtain as perfect a match as possible unlocked many of the secrets of the HLA system. It wasn't long before researchers began to notice that victims of certain diseases seemed to have some of the same HLA antigens in common.

The genes that specify the HLA antigens are located next to one another on the sixth chromosome in humans. There are four points where the HLA antigens' genetic material resides on the chromosome. Each point is called a locus. They have been designated A, B, C, and D. As is the case with every chromosome, there are two versions of chromosome 6 that form a homologous pair. Thus, each person has a locus at identical points on each chromosome that determine the production of HLA antigens. When someone's blood is tested, doctors are likely to find evidence of many different antigens. The list is rapidly growing, but HLA experts now count dozens of possible antigens.

When sperm and egg are formed, the same thing happens with the HLA antigen genes that occurs with any other genes. Half of the sperm or eggs will have one particular combination of the four HLA antigens, the other half will have another combination. When fertilization occurs, the sperm provides half of the new individual's complement of HLA antigen genes and the egg provides the other half, so that the new individual now has eight. The HLA antigen genes are in line on chromosome 6, but researchers don't know exactly how close they are to each other. There may be other genes in between.

Nicholson's patients began coming to the Columbia-Presbyterian Medical Center to have blood drawn for the HLA studies in 1975. It wasn't long before Nicole Suciu-Foca and her colleagues could see clearly that the W3 antigen found on the D locus was common to the diabetic children in each family under study. The obvious implication was that a gene located near the HLA D locus was linked to the presence of the DW3 antigen and somehow was involved in causing juvenile diabetes.

Nicole Suciu-Foca is Romanian. She smokes cigarettes, holding them between her thumb and forefinger in a manner I have always regarded as "European." I met her in her laboratory at the Columbia-Presbyterian Medical Center. She is in charge of a large staff of students, fellows, and scientists, and she obviously follows a hectic pace. As we made our way from the laboratory and down a hall to her office, she was continually interrupted by subordinates with questions about problems with their experiments. Once we were closeted in her office, the telephone constantly demanded her attention. I asked her about her unusual accent.

"The first language I spoke was German," she explained. "The next I spoke was French. Then when I went to school I began speaking Romanian. I grew up in Romania, but the bourgeois in Romania spoke French and German before they even began considering Romanian.

"Just about everyone in Romania decides they want to defect from that country as soon as they begin thinking. I think I was only a child when I began dreaming of defecting. But you have to be careful to pick your time. I had two chances before I actually did, but I passed them up because my husband was still in Romania each time. They only let you out of the country to attend a scientific meeting or something like that if some of your family remains behind.

"In 1969 we worked it a little differently. I won an international scientific prize, a fellowship to study in the U.S. Winning that prize was a big glory for me and for Romania. They had to let me come. When they issue you a passport, that means you are supposed to return. If you don't, you are judged for high treason. But I knew that the day they issued me a passport, if my husband and children were near, I would defect.

"They let me out and then they let my husband, who is a surgeon, out too. Our children were still there and they thought that was enough to make us return. But we calculated differently. We gambled. We thought that if we defected we could still find a way to get our children out. It was a calculated risk. But it worked. I even got my mother out. I managed to get them all out, one by one, with a lot of pressure. My position in this country and my international reputation as a scientist helped.

"When I saw what it is like in this country to be a scientist, I knew I had to stay here. There is no question that I would never have accomplished as much as I have if I hadn't been working as a scientist in this country. The U.S. is the scientist's dream. The Americans have the computers and all the equipment and the money. You really must be born to be a good scientist and I feel that I was born to it. My best ideas came to me in Romania. By Romanian standards I had a superb lab. There were twelve technicians, but no machines. Nothing but microscopes and centrifuges. Those were not sufficient for the research I wanted to do. In Romania there was not enough money to permit me to do pure research. I would have to be involved in clinical research. I didn't want to do that. It would be like a great mathematician having to do computer work.

"But in this country, I could do basic research. That is my addiction. To be a basic scientist is a superb profession. You cannot imagine the excitement, the thrill, when you discover something different or unique.

"The O'Casey family was like that for me when we typed their HLA antigens. You must understand first about genotypes and phenotypes. The genotype is the genes that are actually present on the chromosome. The phenotype is what the genes express, that which we can see or detect through our various tests. When you type a blood sample for HLA antigens, you can see certain phenotypes, but that doesn't immediately reveal the genotypes. To learn what the genotypes are you must resort to other, more sophisticated tests.

"There is great diversity among human beings. It is possible for there to be about 35 million different phenotypes. But when I first typed the O'Casey family's blood samples, every family member was phenotypically identical at the HLA A locus. Each one had an HLA A1 antigen and an HLA A28 antigen.

"At first I couldn't believe my eyes. So I repeated the tests. I got the same results. I knew I had to use more sophisticated tests to unravel their genotypes. I began reacting the blood cells of each family member against the cells of other family members.

By this method I eventually unraveled their HLA genotypes. It was quite astonishing."

As Suciu-Foca talked about the O'Casey family's HLA antigens she grew visibly more excited. She puffed agitatedly at her cigarette. She tore a large piece of paper from a computer printout sitting on her cluttered desk and drew a diagram, which looked like this:

	FATHER			MOTHER	
A_{28}	BW_4	DW_1	A_1	B_8	DW_3
A_1	B_8	DW_3	A_{28}	B_8	DW_3

"What is so extraordinary is that Frank and Mary O'Casey are almost identical at all four HLA antigen loci. Except for Frank's BW_4 and DW_1, they would have been identical. They both are descendants of Anglo-Saxons in Ireland, but for two people from different families in which there had been no intermarriage to be so similar was highly unusual, to say the least.

"From the other studies of diabetic families we knew that the presence of the DW_3 antigen is associated with diabetes. The suspicion was that this antigen on the D locus was associated somehow with diabetes. Probably it meant a gene was present which predisposed someone to develop diabetes. To get juvenile-onset diabetes a child would not only have to have the gene, but something in its environment would be necessary to trigger the diabetes.

"The picture was equally exciting when we determined the three O'Casey girls' HLA genotypes."

She drew another diagram.

	CYNTHIA			MARGARET			CATHERINE	
A_{28}	BW_4	DW_1	A_{28}	BW_4	DW_3	A_1	B_8	DW_3
A_1	B_8	DW_3	A_1	B_8	DW_3	A_{28}	B_8	DW_3

"When I saw the results of the three girls' genotypes, I was amazed. The striking thing is that Margaret and Catherine are identical for DW_3. They each have a double dose of that antigen. And we already knew that DW_3 was related to diabetes.

Catherine had diabetes, and the double dose seemed to have something to do with it. But Margaret also had a double dose of DW_3. She didn't have diabetes. Did this mean she would develop diabetes? Did this mean it might be possible to type the HLA antigens of a child and predict, based on the presence of DW_3, that they might be more susceptible to getting juvenile-onset diabetes than someone else?

"Before I could think about those questions, I had to make certain that Margaret and Catherine *were* identical. For them to be identical would require that a crossover occur when the father's sperm was formed."

Suciu-Foca explained that crossovers can occur during the rearrangement of genetic material that takes place as sperm and egg are formed. The chromosomes line up, each chromosome paired with the other member of a homologous pair. Then they entwine and exchange parts. This is called crossing over. Normally, all of Frank's sperm would have either of two possible arrangements of HLA antigens. Suciu-Foca drew another diagram.

FATHER'S SPERM POSSIBILITIES

A_{28} BW_4 DW_1 A_1 $B8$ DW_3

ADDITIONAL POSSIBILITIES AFTER CROSSOVER

A_{28} BW_4 DW_3 A_1 $B8$ DW_1

MARGARET'S CONCEPTION

Sperm   Egg
A_{28} BW_4 DW_3 A_1 $B8$ DW_3

MARGARET'S RESULTING GENOTYPE
(Identical at D Locus to Catherine)

A_{28} BW_4 DW_3

A_1 $B8$ DW_3

"I became convinced that a crossover on the paternal side had occurred in Margaret's case and that the two girls were apparently identical for the HLA D locus. If Catherine had diabetes, why didn't Margaret?

"Some of the research into the genetics of diabetes has involved twins—both identical twins and fraternal twins. The identical twins are identical genetically. But the research has found that in only 50 percent of the cases where one identical twin develops juvenile diabetes does the other twin also develop it. So you can say right away that if one sibling has diabetes and the two are identical for DW3, then perhaps there is a 50 percent chance that the other child will develop the disease.

"There was another question facing us. Mrs. O'Casey has a double dose of the DW3 antigen gene. But did that mean that a linked diabetes gene was on each chromosome next to each DW3 gene? It was possible that one DW3 wasn't linked to a diabetes gene. If that were the case, could the two DW3 genes that Margaret and Catherine each have be different? You assumed Catherine had a double dose of DW3 linked to the diabetes gene. But because Margaret didn't have diabetes, maybe one of her DW3 loci was different. If they were different, you could make a case for a crossover also occurring in Mrs. O'Casey's egg when Margaret was conceived and giving her, not two identical DW3 genes in terms of linkage to diabetes, but one DW3 linked to diabetes and one not linked to diabetes.

"You must remember that Catherine had gotten diabetes when she was three years old. At the time of our studies, the child with diabetes was seven and her HLA D identical sibling was twelve. Did that mean that Margaret had escaped the environmental factor, whatever it might be? Personally, I had the feeling that she was still in a very high-risk category for developing juvenile diabetes, although I was wishing I could show that there was a real HLA difference between the two girls."

Suciu-Foca paused to light another cigarette, and I asked the question that had been nagging at me for several minutes. "If you could see that Catherine, who had diabetes, was identical to Margaret, who didn't, did you tell the O'Caseys that their daughter was at risk for diabetes?"

She expelled a long stream of smoke and regarded me through it. "No, I didn't tell them," she said after a moment. "You see, I believe you have to have good and very sound grounds to get somebody worried like that. If you can reassure somebody, you should do that. So I expressed to the family my hope that the kids are different, although I couldn't prove it. I felt at that point hoping is better than becoming discouraged, particularly when you don't have conclusive evidence. I kind of said to myself, 'If I am right, she will get diabetes. If I am wrong, then she should not.'"

"I saw Margaret's diabetes coming, but I didn't let myself think about it," Mrs. O'Casey said. As I watched her speak, I was struck by the emotions that seemed to be passing over her face. I had a feeling that it was painful for her to utter each word.

"For about two weeks, when Margaret would come home from school, I would notice how she would come flying up the stairs. You could hear her. She would come through that door and race into the bathroom. Something struck me, but I didn't want to think about it. I put it out of my head. Then one morning about 6 A.M. she came into our bedroom and said, 'Mommy, I can't hold it.' She had this sensation, this pressing, of wanting to urinate and not being able to hold it.

"Also, she was irritable, very cranky. We thought her crankiness was from her schoolwork. She tries very hard in school and she had hard courses then. So I made up excuses. One day I said, 'Do you have trouble with urinating anymore?' She told me no. But I said, 'Let me test your urine.' Well, I tested her and the sugar in her urine was sky-high. I knew right then she had diabetes. I called Dr. Nicholson and he said, 'Bring her right down.'

"We went to Columbia-Presbyterian to have a blood test. I didn't see Dr. Nicholson when we arrived, but I spoke to his secretary. She sent us off to have the blood drawn. When we finished, they said it would be about thirty minutes or an hour. I asked the secretary what we should do and she said, 'Why don't you go home? There's no sense waiting around here.' So we left the hospital and I think Margaret thought she was free and clear, that she wasn't diagnosed as a diabetic."

Frank O'Casey interrupted. "While they were on their way home, the hospital called and said there had been a mistake. She was to return immediately. Her blood sugar count was very high. It was a shock to her when she got home. We had to say, 'Sorry, Margaret. You have to go back.'"

Mrs. O'Casey continued. "When I think about it now, perhaps

God in his wisdom wanted her to come home, so we could be in the house with her. She could let her emotions out. If we had stayed in the hospital, perhaps that would not have been the best thing. So we had it all out. We cried and cried. We took a walk, a long walk. We went shopping. We got away from it all. Then late in the afternoon we went back to the hospital and they admitted her. I think being at home helped her accept that she is a diabetic. She's had a much harder time of it than her little sister. Catherine can roll with the punches more. Margaret can't. Perhaps it has something to do with her age. Catherine was three when she got diabetes. Margaret was twelve."

So Nicole Suciu-Foca had been correct. Margaret apparently was HLA-identical to her sister. Something in their environment had struck them, obviously several years apart. What could it be? I asked John Nicholson.

"The popular idea is that a virus infects the genetically predisposed child and attacks the beta cells in the pancreas," he said. He had finished his coffee and paused to light a pipe. "There is evidence linking mumps virus and a lesser-known virus called Coxsackie B4. It has been shown that outbreaks of juvenile diabetes in some communities follow three or four years behind mumps epidemics.

"My personal view is that a virus is involved and that it could be more than one virus. The question is, how does the virus affect the beta cells? There are three possibilities.

"The first is that the virus infects the body and causes the immune system to manufacture antibodies to the virus. The antibodies attack the virus. But somehow a mistake is made in the production of the antibodies and they are also sensitized to attack the beta cells. What you have is an immune response that cross-reacts with the beta cells.

"Another possibility is that the body has an inability to defend itself against the specific virus that attacks the beta cells.

"The third idea is that an immune response is mounted against the attacking virus but for some reason it isn't turned off when the virus is gone. This same immune response then cross-reacts and goes after the beta cells. I prefer this last idea because

I don't think kids with juvenile diabetes are more susceptible to viral infections than other children.

"If the immune system is involved in responding to a virus and somehow causing destruction of beta cells, then you begin to think of ways to intervene and prevent damage to beta cells. You want to find some component of the immune system that responds to the virus but doesn't cross-react with the beta cells. If you identify that, then you might be able to immunize a child. That way, when the virus appears, the immune response caused by your immunization would attack the virus and leave the beta cells alone.

"I think we're going to see progress in this direction."

I first met the O'Caseys in July, just three months before Margaret's diabetes was diagnosed. Riding toward midtown Manhattan on the A train, the same train the O'Caseys take each time they visit Columbia-Presbyterian, I sat in a swaying car jotting down my impressions. "See them again when the children aren't present," I wrote. There was more complexity to this couple than they had revealed. They were friendly and cooperative, but they held something back.

I returned late the following October. Speaking to Frank O'Casey on the telephone, I candidly told him I wanted to speak to him and his wife while the children were out. We selected a time when the girls would be in school.

But when I arrived, Cynthia was unexpectedly at home. She was in the bedroom that is separated from the living room by french doors. One door was ajar, and I could hear her moving about. Mary hadn't come home yet from her part-time job. I plunged ahead and asked Frank if he felt guilt about his daughters' disease and the genetic factor involved. I could sense him struggling to formulate an answer.

"At first, when we found out about the youngest coming down with diabetes, I had guilt feelings, suspecting that I passed it on. That was one of the first things we heard from somebody, a nurse or someone, that it was genetic. Later, we found out we are carriers and did pass it on. But there was nothing could be done about it."

If he had known about the possibility of diabetes in advance, would he have changed anything, I asked.

"I don't know," he said. "I've never gone that far into it. It's . . . it happened. And that's that."

Did he ever feel bitterness, anger? I could see him watching the door into the bedroom. The daughter behind that door had inherited his "safe" DW1 HLA antigen gene. She had escaped.

"No. No bitterness," he said slowly. "Everybody has problems in life. It's hard to accept, but you help one another. You counsel each other."

The bell for the electric lock on the apartment building door downstairs buzzed at that moment. "There's my wife now," he said, rising. He was probably glad to have her arrival divert my nagging questions.

Mrs. O'Casey bustled into the room a few minutes later and gave me a warm welcome. She sat by her husband on the couch. I told her that I wanted to talk about guilt feelings. Her face creased into the same painful look I remembered from three months before. But she pushed gamely ahead to answer my questions.

"In the beginning, it was very difficult to accept," she began. "There was a certain amount of guilt on my part that I was the cause of this. Somehow or other you learn to come to grips with it. When we were married we didn't know we had any such thing. If I'd known, who knows what we would have done. But you can't change things. You can't go back. But because I didn't know, I just had to erase the guilt feelings. And I don't think they have ever cropped up again."

Margaret came home from school then and passed through the living room into the bedroom. She smiled at me and cast an uneasy look at her parents. I felt my presence embarrassed her, perhaps reminded her of things she would like to forget. The door was still ajar.

When Mrs. O'Casey resumed speaking, she began to lower her voice. As she spoke, probably without realizing it, she lowered it even more until she was almost whispering.

I told her about Nicole Suciu-Foca's desire to test all Mary's brothers and sisters and her parents. Mary reported that her sister in Maryland had traveled to New York with her children to have the blood tests done. They hadn't heard what the results were. I emphasized how important Suciu-Foca considered obtaining blood samples from Mary's mother and father.

"My mother has more guilt feelings about this than anyone," Mary said. She glanced at the bedroom door and lowered her voice even more. I strained to hear. "Yes, she does," Mary said, glancing at her husband as if she expected him to dispute her. "She is that type. Evidently I got one gene from her and one from my father. My mother is definitely afraid. They're both up in their seventies and maybe they should be left alone. They've had enough in their lives. But I don't see how it could really bother them that much. I think they might come in and give blood if somebody else got to them . . . instead of me. But my mother's definitely afraid. She doesn't want to know."

I left a few minutes later, feeling frustrated. I sensed pent-up feelings. As a family, they seemed close on the surface, but distant underneath.

Outside, it was unseasonably warm for a late October day in New York. I sat down on a bench beside a park, surrounded by elderly people bundled in coats taking in the warmth from the sun. It was pleasant to listen to the musical sound of the Irish accents.

As I pondered what Frank and Mary had said, I decided it wasn't Mrs. O'Casey's mother who was afraid to find out as much as it was Mrs. O'Casey herself.

SANDY

The number of chromosomes found in the nucleus of a cell is a fundamental biological trait that distinguishes one species from another.

Although chromosomes were discovered before the turn of the century and biologists were convinced they contained a cell's genetic material, it wasn't until 1944 that scientists proved this genetic material is DNA. They showed that a nonvirulent strain of pneumococcus could be made virulent by transferring some chromosomal material from one bacterium to another.

Chromosomes can be seen with a microscope. Studying human chromosomes was difficult because they are jumbled within the cell's nucleus and look alike. It was generally thought that humans had 48 chromosomes, which existed in 24 similar or homologous pairs.

Then in 1956 Joe Hin Tjio and Albert Levan, working at the Institute of Genetics in Lund, Sweden, discovered there are only 46 chromosomes and they exist as 23 homologous pairs. Their finding caused considerable excitement among the biologists interested in such matters. Two developments made their discovery possible. One was treating cells with a solution lower in salt content than the interior of the cell, causing the cell to take up water and swell, spreading the chromosomes farther apart. The other was discovering that colchicine, a substance derived from the autumn crocus, can halt division or replication of a cell at a point where the chromosomes are particularly visible. Armed with these new techniques, biologists could see human chromosomes well enough to divide them into seven groups according to size and location of the centromere, a conspicuous constriction in each of the rodlike structures.

Interest in cytogenetics, the study of chromosomes, burgeoned in the 1960s, but it was a frustrating field of research. Most chromosomes could be identified only as belonging to one of the seven different groups. Only a few could be individually identified.

Then in 1970 a fluorescent dye called quinacrine was used to stain the chromosomes, permitting specific identification of each one. The dye caused the chromosomes to appear to be striped or banded, with enough differences among them to make it possible to identify not only each chromosome, but each part of every chromosome. A number of additional banding techniques were discovered quickly after that, including Giemsa banding, trypsin-Giemsa banding, reverse banding, and C banding. The long-sought ability to see the chromosomes in considerable detail made studying genetic diseases caused by chromosome abnormalities one of the fastest-growing branches of medical genetics in the 1970s.

Geneticists now know of several hundred chromosome anomalies. Most are one-time occurrences that take place during the formation of sperm or egg and produce an affected fetus. Others can be passed from generation to generation within a family, but are unique to that family. A third but quite rare group is passed from generation to generation and seen in many different families.

One of the best-known chromosome anomalies is trisomy 21, which causes Down's syndrome, or mongolism. In these cases, there is an extra chromosome 21. The anomaly arises when a woman produces an abnormal egg; more rarely, when a man produces an abnormal sperm. It is seen most frequently in the children of women over thirty-five. Down's syndrome victims seldom reproduce, but when they do the chromosome defect is transmitted to their offspring.

Chromosome anomalies are of two types: those caused by an alteration in the number of chromosomes, and those caused by an alteration of chromosome structure produced by breaks and rearrangements. These can take several forms. A piece of chromosome can be broken off and lost; a piece may become inverted; or both tips may be lopped off and the raw ends join to form a "ring" chromosome.

Sometimes more than one chromosome is involved. A piece of one chromosome may break off and insert itself into another. This is called a translocation. Geneticists speak of balanced and unbalanced translocations, depending upon whether the breakage and rearrangement results merely in the genetic material being moved about or in an increase or decrease in the total genetic material normally present.

It was a translocation, passed from generation to generation and traced back to her ancestors in the village of Malisowisna in Poland, that brought Sandy to the attention of the doctors at the University of California at San Francisco. She had become pregnant and wanted to determine if her baby might be born retarded like her two older sisters. As the result of an unfortunate breakdown in communications about Sandy's case in the summer of 1976, the young woman became the center of an urgent, last-minute effort to unlock the secret of her chromosomes. The UCSF medical team anxiously sought an answer before it was too late for Sandy to end the pregnancy if she desired.

At the end of the sixth-floor wing of Moffitt Hospital where the UCSF medical genetics group has some of its offices is a conference room with a blue carpet, an oblong table surrounded by chairs, a blackboard, and rows of folding chairs against the wall opposite the blackboard.

Every Monday the genetics group meets during the noon hour to discuss current cases and prepare for the weekly genetics clinic the next day. The people who straggle into the conference room on Mondays bring their lunch—a carton of yogurt, a paper bag holding a sandwich and an apple, cups of tea or coffee they've brewed in their offices, or a sandwich and bag of potato chips from a vending machine. They sit around the table or on the folding chairs, and after a few minutes of eating and bantering conversation the meeting gets under way.

Besides the attending physicians in the genetics group, the full-time genetic counselors and technicians, and the head of the cytogenetics laboratory, a constantly changing group of postgraduate medical fellows and interns and residents rotating through the genetics program attends the meeting. The University of California at Berkeley across San Francisco Bay has a program to train medical genetics associates, a new group of ancillary health personnel who assist physicians in dealing with genetics cases. These students gain their clinical experience with the Moffitt Hospital group and always attend the Monday meetings.

Sandy's case came up for discussion at a Monday meeting in July of 1976. It was quickly obvious that some of the physicians around the table were in disagreement about how to handle her case. To those who knew him well, it also was obvious that Charlie Epstein, the doctor who directs the UCSF program and moderates the Monday sessions, was becoming irritated with the situation.

Sandy was scheduled to appear the following Monday for an amniocentesis to determine if the chromosome anomaly that had been found in her, her mother, and her two sisters was present in the sixteen-week fetus she carried. The discussion, which really couldn't be called an argument, centered upon whether the amniocentesis should be performed and whether it would reveal anything.

At one end of the table, Mickey Golbus, the obstetrician-geneticist who performs most of UCSF's amniocenteses, was arguing against the procedure. (He is the same doctor who collaborated with Gloria's doctors at Yale.) A Chicago native and then thirty-seven years old, Golbus was educated at the University of Illinois school of medicine and came to California for his postgraduate medical training. He jokes with friends that he is the Jewish mother's delight: a son who became a doctor. Mrs. Golbus, in fact, is thrice blessed with three doctor sons. As Golbus spoke, he munched on a sandwich that he had brought from his home in Marin County, and sipped tea from a handmade ceramic mug.

"Sure we can do a tap," he said earnestly. "But what's it going to tell us? It's not going to tell us enough to justify the procedure. All we know about this family is that they have extra material on the long arm of chromosome 16. But the mother and daughter who are normal have it and the two daughters who are affected have it. So what can we tell her about her baby if we see a long arm of 16?"

"I disagree, Mickey. We can tell her something," answered Bryan Hall, an assistant professor of pediatrics and one of the medical genetics group's physicians. A tall, heavy man who is a native of Kentucky, where he received his medical education, Hall speaks with a rich southern drawl. He entered the Air Force after finishing medical training and spent his time at Travis Air Force Base near San Francisco. An interest in genetics brought him to UCSF and a position on the faculty after his military service and a fellowship in Seattle were completed.

"What we can tell them, Mickey, is whether or not chromosome 16 appears normal. If it's normal, then we know the fetus is unaffected. If we find a long arm on 16, then we can tell her there is a 50 percent chance the baby will still be normal."

Hall and Cyndy Curry, a young doctor completing a genetics fellowship with the UCSF group, had been more involved with Sandy's case than the others. But three weeks before, Curry's fellowship ended and she left UCSF for another position. She wasn't present to support Hall in the discussions. It was Cyndy who had seen Sandy five weeks earlier in UCSF's satellite genetics clinic at Santa Rosa in Sonoma County north of the city. She learned Sandy was pregnant and counseled her to come in for an amniocentesis. Now Golbus was arguing that the amniocentesis wouldn't be useful.

As the two physicians talked, Charlie Epstein leafed through Sandy's file. He hadn't been involved in the case until now. He studied her records, noting what the cytogenetics tests had shown and what his colleagues had told the young woman and her family at prior counseling sessions. Epstein idly ran a hand through his tousled brown hair flecked with gray. He wore glasses with metal frames and had on a sports shirt and slacks. He frowned as he read the reports.

"I think Mickey's right," he said finally. "You can't tell them enough to justify an amniocentesis. If there's extra chromatin on the 16 chromosome, you won't know whether the fetus is just a carrier or is affected. And worse than that, even if all of the chromosomes look normal, we won't know whether one of them has a tiny deletion we can't see." He turned to a man in green slacks who sat in the front row on one of the folding chairs. "What do you think, Bill?" he asked.

Bill Loughman, a man of middle age with a gray-flecked

beard and a mustache whose tips had been carefully waxed and turned upward, is head of UCSF's cytogenetics laboratory. He and his assistants perform the delicate tests on cells grown in culture that make chromosomes visible for study. Loughman is not an M.D. but a Ph.D. in genetics, a man who worked for fifteen years at the Lawrence Radiation Laboratory of the University of California at Berkeley and earned his master's and doctorate degrees during that period, one course per semester. He had recently been hired to expand and reorganize the university's cytogenetics laboratory and was wrestling with equipment and space problems and dissension among some staff members unhappy with the changes he was making.

To a certain extent, the problem facing the genetics group that day was an affront to his professional pride. The question that his laboratory couldn't answer was where the extra piece of material on chromosome 16 in Sandy's family came from. If only his laboratory's procedures were sensitive enough to see all the chromosomes in sufficient detail to answer that question, then the discussion wouldn't be taking place. Apparently there was a translocation involved, but their tests wouldn't reveal whether it was balanced or unbalanced in any particular individual.

Loughman shook his head at Epstein's question. They seemed to have exhausted the laboratory's bag of tricks.

"Perhaps what we should do is this," Epstein said, seeking a compromise. "When the patient and her family come next Monday, Mickey won't do an amniocentesis but we will draw more blood, prepare new slides, and send them to Uta Francke in San Diego. We'll tell her it's urgent and see if she can see anything. We should know within a few days. If she can, then we would still have time for the amniocentesis."

Heads nodded in agreement. Uta Francke is one of the finest cytogeneticists in the country. Maybe her laboratory could see something in Sandy's elusive chromosomes.

"Now let's move on to the next case or we'll be here all day," Epstein said, glancing at his watch.

Charlie Epstein entered medical genetics through the back door. He went to Harvard College and Harvard Medical School and then did his internship and residency in Boston hospitals.

Like so many of the best people in academic medicine, he secured an appointment as a research scientist at the National Institutes of Health in Bethesda, Maryland, as a means to satisfy a military obligation. He spent two years investigating the structure of proteins and then was sent by NIH to the University of Washington to study genetics for a year. He returned to NIH and spent three more years in protein structure research. He started a genetics clinic on the side at NIH, but it was never successful. He began searching for an academic appointment in genetics and in 1967 moved to the University of California at San Francisco to establish a medical genetics program.

"What I remember most about the day we discussed Sandy's amniocentesis was my emotional reaction to it," Epstein told me later. "I was really shocked that we had gotten ourselves into that kind of bind, where we had a patient thinking everything was all right and that we could give her answers, and all of a sudden finding that we didn't know where we were. I was upset that somehow communications had gotten fuzzy enough so that she could get to the point of being pregnant and coming in to have an amniocentesis and being told there was nothing we could do. I hadn't been involved in the case until then, but I could see we were in trouble. We had about a week or ten days to come up with an answer.

"We have an excellent cytogenetics lab, but sometimes another laboratory can see something that your own lab can't spot. It's a difference in procedures and techniques. Everyone knows Uta Francke by her reputation. It was worth sending new slides to her."

One of Epstein's accomplishments during his decade at UCSF was establishing a system of satellite genetics clinics to serve outlying regions of northern and central California. Members of the UCSF genetics group make visits every two or three months to the satellites, operated in conjunction with county health departments, and offer genetic counseling to local residents.

One of the clinics is in Santa Rosa, a town nestled among the green hills of Sonoma County fifty miles north of San Francisco and just over a range of hills from the Napa Valley where California's most famous wines are produced. It was through the Santa Rosa satellite that the chromosome abnormality affecting

Sandy, her mother, and her two retarded sisters first came to the geneticists' attention.

When they married in San Francisco in 1951, Don and Cecelia, to whom I'll give the last name Muller, knew in a vague way that something was "wrong" with Cecelia's family. Her younger sister, Dolores, was retarded. Back in Milwaukee, where Cecelia grew up, her Aunt Josephia had two retarded sons, Edward and Casimir. Aunt Stanishawa's daughter, Wanda, who also lived in Milwaukee, was retarded. They had gone to special schools as children. Some were eventually placed in institutions. Others seemed able to function in a limited way at home. As a teenager, Cecelia's sister Lorraine vowed never to marry and have children because of the fear of producing babies like their sister Dolores and their cousins. She did marry later. "He was irresistible," she now jokes about her husband.

Cecelia and Don and Cecelia's family knew nothing of genetics, but it was obvious something was passed from generation to generation in the family. But it didn't deter them. They set about raising a family, and whatever doubts might have existed were relegated to some faraway drawer in the subconscious.

Don and Cecelia had three daughters in rapid succession, starting in 1953: Suzanne, Stephanie, and Sandra. Don and Cecilia soon began to wonder if something was wrong with Suzanne. She seemed slow. She was small. One eye didn't point correctly and required corrective surgery. She was unusually afflicted with dental cavities. Their second child, Stephanie, seemed slow, too. She had double hernia surgery as a toddler and also was highly prone to cavities. But it wasn't until each daughter entered kindergarten and was observed by teachers and tested that the mental retardation was confirmed. Their two oldest daughters were classed as "educable." Sandy, to their relief, was normal.

Cecelia had come to California from Milwaukee as a young woman because her sister Lorraine was already on the West Coast. Lorraine married at about the same time and moved to San Diego. Lorraine gave birth to a son, Jeff, just a few weeks after Don and Cecelia's second child, Stephanie, was born. Jeff had a double hernia operation when he was only eight months

old. He developed a hip problem that required wearing casts for
nearly eight years. He died of pneumonia at age twenty-one, and
physicians thought a cardiac arrhythmia or erratic heartbeat
might have contributed to his death. Lorraine and her husband
attributed Jeff's small size to his orthopedic problems, which at
one point required nearly a year of hospitalization. It wasn't
until the second grade that his retardation was confirmed.

Lorraine's second child, Eileen, was normal. But her third
child, Blaise, was classed as retarded in school and attended special classes. A fourth child, Mary Ann, was normal.

Finally, Cecelia's cousin Helene, who lived in Milwaukee, had
three children. The last child, Sparky, who was born not long before Don and Cecelia's first child, was classed as retarded when
he entered school.

By the time Sandy was ten years old, there was no doubt in
the minds of her parents and aunts and uncles that the family's
"problem" had been visited upon another generation. Five retarded children had been born.

"As kids growing up in Milwaukee, none of us ever discussed
this much or thought about it. Our parents didn't question it, at
least not openly," Cecelia told me when I went to Santa Rosa to
talk to her and Don and visit Sandy and her husband. I spoke to
Don and Cecelia in the breakfast nook of their rambling single-
story stucco house in an older residential section of the city one
morning as torrential rains fell outside, relieving a two-year
drought that had plagued northern California. Cecelia is a large,
jolly woman and Don is short and wiry.

"I think as we grew older some of us wondered about it,"
Cecelia said. "I know my sister Lorraine did after Jeff and Blaise
were born. Helene back in Milwaukee questioned it when she
realized that Sparky was retarded. But when we were growing
up my sister and our cousins went to special schools. We went to
our schools, they went to theirs. My sister Dolores could ride the
streetcar to school and do simple things.

"Lorraine was the first to do something about it, such as going
to a doctor. She took Jeff and Blaise when Jeff was about eight
up to Children's Hospital in Los Angeles, and a doctor there examined them. A group of doctors was doing some kind of re-

search. That was in about 1964. They looked at their chromosomes and couldn't see anything. I think they might have said
the 16 chromosome had something extra, but they said that chromosome sometimes has extra stuff anyway. So they really
couldn't say what was wrong.

"After Sandy was born in 1955 I got pregnant again, but I
bled constantly. I went to the doctor about it and he gave me
some medicine and said if the bleeding continued I should call
him. The bleeding didn't stop and finally the baby came out. I
had a miscarriage. It was just a mass of tissue. I saved it and
gave it to the doctor in case he wanted it for study.

"Then about 1957 or 1958, when we knew that Suzanne was
retarded and then that Stephie was the same way, I became
pregnant again."

Don broke in. "She's a very strict Catholic. But I said I didn't
want another one."

"Yes," she said. "We decided since two of our three kids were
retarded, we just didn't want to take the chance on having another one. So we went to our family doctor and he sent us to
other doctors and you had to get three doctors together then to
okay the abortion. I think that was the law. So they agreed and
they took the baby."

"I was all for it then," Don said. "But now I think maybe I
made a big mistake. She should have had the baby."

"I think I feel bad about that, too," she agreed. "I always felt
bad about that."

"I was really tough on her about it," Don continued. "The
main reason was that I lived a hard life. I had no father. Some of
the things I wanted from life I never got because I had a mother
to support and then World War II. So I felt I already was tied
down by two retarded children and I didn't want another. But I
always had a guilty conscience after that. I should have said,
'Okay, we'll have this one and then no more.'"

"When Lorraine's son Jeff died in 1974, it had an effect on all
of us," Cecelia continued. "That started me to thinking about it
some more. And then Sandy got married not long after that. At
the end of 1975 when she and her husband, Joe, were thinking
about having children, they decided to go and ask a doctor here
in Santa Rosa about it. Sandy told him about her retarded sisters

and asked what the chances were of something like that happening to her. And the doctor said he didn't know anything about it but he could put Sandy in touch with someone who did.

"So it wasn't too long before Joan Weetman, who is a public health nurse here, called Sandy at home and asked her if she could come over and talk to her about the problem. Mrs. Weetman is connected with Epstein's people in San Francisco. She went over to Sandy's one night and she started trying to put together a pedigree. Sandy couldn't answer a lot of the questions, so she called me and I went over. Then a few weeks after that, we all went in when the doctors from San Francisco were here for the genetics clinic. We took Suzanne and Stephie, and Sandy went along. Dr. Hall and Dr. Curry were there. And that's how it all started."

"As a group the chromosomal disorders affect roughly one in 240 births and are a major category of genetic disease," Epstein told me. "They affect a much larger number of conceptions. There are various estimates about spontaneous abortions during the first three months of pregnancy, but at least one half of them are caused by chromosomal anomalies. So if you figure about 15 percent of all known conceptions are aborted in the first trimester, then that means at least 7.5 percent of all conceptions result in abnormal chromosomes. They probably represent the most common defect in terms of total conceptions among the population.

"You can see that the majority of the chromosome anomalies are weeded out through spontaneous abortion during the early weeks of a pregnancy. The ones that survive may seem severe to us looking at them, but they are much less severe in terms of development than the ones that don't make it.

"There are many things that can cause chromosome anomalies. Some of the well-known ones include radiation and certain drugs. In Down's syndrome the mother's advancing age is often a cause. But sometimes a particular chromosome anomaly will run in a family. Somewhere in the family's past, probably in one person, there was a chromosome accident in which a piece of one chromosome was shifted to another. Sometimes it's a large piece, sometimes a small one. A normal individual has two copies of each gene. But depending on how things sort out with a chromosome anomaly, you can have a situation in which the offspring have three copies of a group of genes. We call that a trisomy. Or there can be one copy of some genes. That's a monosomy. As a rule the trisomies are less severe than monosomies. A person with a partial trisomy may be retarded and have physical abnormalities but often survives. A monosomy usually results in a spontaneous abortion.

"We know there are families that can pass along a translocation. The family members who have a balanced translocation are carriers. Some of their children are normal, some receive a balanced translocation and are carriers, and some have an unbalanced translocation and are affected. They are the ones with a trisomy and are often retarded or have physical abnormalities. You often see frequent miscarriages among the carriers, and the presumption is those are the monosomics. These conditions aren't transmitted by classical genetic means. You frequently see erratic patterns. It can be very puzzling—and very challenging."

Challenging is how Bryan Hall and Cyndy Curry regarded Sandy's case when they first saw her on December 11, 1975, at the Santa Rosa clinic. Only Cecelia and Sandy came the first time. The two doctors questioned them in detail, filling in pieces missing from the initial work-up Joan Weetman had performed at Sandy's home. They were particularly interested in the medical histories of Cecelia's relatives. After mother and daughter left, Hall and Curry spent nearly forty minutes at the blackboard in a room of the Sonoma County Health Department building drawing the family pedigree and debating its meaning. The patterns of inheritance didn't make sense. The numbers didn't fit. But slowly, by a process of elimination, they convinced themselves they were dealing with a chromosome anomaly, probably a translocation—balanced in Sandy and her mother and unbalanced in Suzanne and Stephanie. The next step would be to examine the two retarded girls and draw blood from everyone for chromosome studies.

It was February 19 before the two doctors and Cecelia, Sandy, Stephanie, and Suzanne got together again in Santa Rosa. The doctors spent considerable time examining Suzanne and Stephanie, taking measurements, snapping pictures, and finally drawing blood from all four.

Hall sent the blood samples to the cytogenetics laboratory the next day. The technicians started cultures of the cells in the blood and then prepared microscope slides for visual examination of the chromosomes when enough cells had been grown. But it was April 7 before the laboratory technologists finally completed their examination of the slides and prepared a report, stating that all four patients had extra material on the q arm—the long arm—of chromosome 16. There was no evidence revealing the origin of the material. If a translocation was present, its site of origin couldn't be seen. There was no way to tell a balanced translocation carrier from an unbalanced translocation causing an affected person.

By then it was too late to schedule Cecelia and her daughter for the April clinic in Santa Rosa to explain the ambiguous findings. The next clinic was planned for early June. An appointment was made for June 10, the last time Cyndy Curry would participate, because her fellowship was ending and she had a new job elsewhere in the state.

Meanwhile, unbeknownst to the San Francisco doctors, Sandy missed her menstrual period, which should have begun the first week in May. It was nearly Memorial Day before a Sonoma County obstetrician confirmed she was pregnant. A few days after that she, her husband, Joe, and her mother met with Curry and Hall in Santa Rosa. Sandy announced her good news.

The doctors were nonplussed. They told their patients about the ambiguous findings with the chromosomes and expressed their disappointment that nothing definitive had resulted. The doctors said they *knew* there must be a difference between Sandy's and her mother's chromosomes and Suzanne's and Stephanie's chromosomes. It was the only way to explain the family pedigree. But they couldn't see the difference. In the end, Hall and Curry told Cecelia and Sandy the same thing the doctors in Los Angeles had told Cecelia's sister Lorraine a decade before—chromosome 16 looked larger than normal, but nothing seemed amiss.

Sandy asked about amniocentesis. They encouraged her. There was a 50 percent chance her baby would have a normal

chromosome 16. Amniocentesis would reveal that. If the procedure showed instead that an abnormal 16 was present, then
there would be a 50 percent chance of a retarded baby and a 50
percent chance of a normal carrier. Having that information,
Hall and Curry agreed, was far superior to not knowing anything about the baby. They explained that Golbus performed
amniocenteses every Monday and Thursday afternoon. They recommended Sandy schedule an appointment for late July. She
agreed. Judy Derstine, the genetics counselor who accompanies
the doctors to Santa Rosa each month, arranged for Sandy to
have her amniocentesis on July 19, a Monday.

Sandy and Joe and Cecelia made plans to drive to San Francisco for the test. Sandy would have the amniocentesis at the
UCSF medical center and return home later that afternoon. But
on July 12 the genetics group met for their Monday noon conference and Sandy's case turned up for discussion because of the
pending amniocentesis. Mickey Golbus opposed performing the
amniocentesis; Hall argued in favor; and Charlie Epstein offered
his compromise. As the discussion progressed, Hall realized that
he and Curry had made a mistake the month before in Santa
Rosa as they counseled Sandy. They hadn't considered that a
third possibility might exist: a deletion that couldn't be seen on
some other chromosome, resulting in a partially monosomic
fetus. Golbus was right. An amniocentesis wouldn't tell them
anything.

On the day Hall and Curry saw Sandy and Joe and Cecelia in
Santa Rosa five weeks before, the doctors had told their patients
there was one more thing that could be done: high-resolution
banding of slides containing their chromosomes. There was a
chance these more sensitive tests might reveal where the extra
material on chromosome 16 came from.

Hall made the request for the additional tests on June 11. The
cytogenetics laboratory subsequently prepared a new set of
slides, treated with a biochemical called trypsin prior to staining
with Giemsa stain. The technique was not in standard use at
UCSF. To the later chagrin of Bill Loughman, the cytogenetics
laboratory's new director, the trypsin-treated slides were lost.
Hall's request was never fulfilled and he didn't follow up to see
what results the extra tests had produced.

"Our ability to recognize chromosomal disorders has changed quite a bit in the last few years," Loughman told me. "The field has been moving particularly rapidly since we have developed the ability to see all the chromosomes and examine them in considerable detail. Now we can look at the tips of chromosomes and other regions that were once hard to see. So chromosome abnormalities that we couldn't diagnose three years ago we can now diagnose, and some we can't diagnose now may be diagnosable two years from now.

"A fundamental development in this field in the 1950s was learning how to cause cells in the blood to grow in cell culture. That provided an easy way to obtain cells to be studied. Then came the stains that opened the field up. What we do is put some of the blood in a culture medium which is filled with various amino acids and nutrients that the cells like. We're dealing with lymphocytes, or white blood cells. We also add a chemical that tweaks them and starts them dividing. They divide for about sixty-eight to seventy-two hours, which gives us enough cells to work with.

"When we have enough cells, we poison the cell culture with a substance that stops each cell dividing at a certain point in the division process, a point we call metaphase. This is when the chromosomes are most visible in the cell nucleus. About 1 percent of the cells are dividing at any one time. So this poison, called a mitotic poison, stops each cell dividing when it comes up to the metaphase stage. We let the culture sit there and collect poisoned cells until maybe 5 percent of the cells have been stopped.

"Then we use a centrifuge to settle the cells out of the culture medium and treat the collection of cells with a hypotonic salt solution, which causes the cells to take up water and swell, become very fat. The chromosomes are just floating in the cell, but the cell is very small, so if you tried to look at it under a microscope, the chromosomes would be all jumbled up. But when the cell takes up water, the chromosomes are spread out.

"After a period in the hypotonic solution, we kill the cells with a fixative which consists of alcohol and acetic acid. Now the dead cells are floating in this fixative. We take a few drops of the solution and put them on a microscope slide. The surface tension of the fixative causes the cells to spread out over the microscope slide and rupture their membranes. The alcohol and acetic acid dries quickly and the cells are left on the slide, dried and relatively stable. It's very much like egg white that dries onto a plate.

"We age the cells on the slides a couple of days, or we use mild heat. This makes them even more stable on the slide. Then come the various treatments that enhance the chromosomes' appearance under a microscope. The technicians are expert at looking through a microscope at a slide and counting all the chromosomes, identifying them by number and looking for abnormalities. They sit there at the microscope mentally lining up the chromosomes and comparing them. They can spot an abnormality almost immediately. Finally, we take a photograph through the microscope of some of the slides and then cut out each individual chromosome as it appears in the picture and paste these snippets of photos onto a sheet so the chromosomes are all arranged in order and you can study them in detail.

"If you can't see all the detail in the chromosomes you want with one stain, then you try another on other slides. There are some stains that are preferred and that most laboratories use. There are others that are more difficult to use and are reserved for special studies. We didn't routinely prepare our slides with trypsin before Giemsa staining. I was still new on the job. Making changes was a slow process.

"I was hired in December of 1975, six months before all this took place. We had four technicians doing chromosome analysis in one location. It was a tiny, crowded room scarcely larger than a broom closet. Another technician was in another part of the medical center doing amniocentesis cultures. And my office was in a third place. There were equipment problems, space problems. Plus we had grown topsy-turvy and were overloaded with over three hundred amniocenteses a year and an equal number of other types of chromosome analyses.

"They hired me to bring it all together, coordinate things, improve the level of output and the quality. But when I came I spent the first six weeks working as a technician because my chief technician went into the hospital ill. When that was over I began trying to reorganize. I tried to introduce new methods.

There was some friction because some of the technicians thought I was to be another technician, which I wasn't. One technician knew how to use trypsin in the staining, but she and some of the other techs thought it was an 'iffy' procedure.

"But I had an extensive experience in the past with trypsin, so in Sandy's case I personally treated a set of slides with trypsin and left them for analysis. And then they were lost. I'm not surprised now that they were. It was a difficult period of too much work, personal clashes, reorganization, with rising work loads and little experience in high-volume cytogenetics work for any of us. Somehow Sandy got lost in the shuffle.

"This is not meant to excuse the loss of the slides so much as to explain it. You can imagine my great chagrin later."

On July 19 Sandy, Joe, and Cecelia drove to San Francisco and met Golbus at the office on the eighth floor of the medical center's ambulatory care center adjacent to the amniocentesis treatment room. Sandy had spent a sleepless night, her fear of the long needle steadily building. To Sandy's relief, Golbus announced that he didn't plan to remove any of the amniotic fluid in which Sandy's baby floated. He carefully explained the reasons why and then asked Sandy and her mother if a technician could take more blood from them. They consented, gave their blood, and returned to Santa Rosa.

"You might ask why we didn't just call Sandy and tell them not to come. But the conversation I needed to have with them was a very sensitive one," Golbus told me. "It's the kind of conversation you just don't hold over the phone. You have to see them in person, sit down with them and explain things. If you have to, you explain things several times until they understand. I wanted her husband to hear it, and her mother, who was the real force in much of this.

"But it wasn't a conversation I was particularly looking forward to. I knew that they had been led to believe that amniocentesis was going to be done and I knew I wasn't going to be able to do it. I was put on the spot of having to tell them. I was perturbed about that. Also I felt some anxiety because you know you're not going to meet people's expectations. I thought it would be a difficult session. But it wasn't. They were great. I explained everything we knew and then had them give blood.

"There was no question the pressure was building by then. At the most we had about ten days or maybe two weeks to get an answer. California law states that an abortion can't be performed past twenty weeks. But it doesn't say whether that's weeks since the last menstrual period, weeks since implantation of the embryo in the uterus, or fetal weeks. They're all different. I don't know if that's an oversight on the legislature's part or whether it was purposely left ambiguous.

"In practice, I've performed abortions up to twenty-two to twenty-four weeks past the last menstrual period.

"So you can sit down and count the weeks and see what kind of pressure we were under. When I saw Sandy on July 19 she was fourteen weeks since her last period. Allow two weeks to get the slides to Uta Francke and for her to come up with results. So she's at sixteen weeks before you could perform an amniocentesis. Then it takes three weeks for the fetal cells you obtain in a tap to grow in the cell culture. Now she's at nineteen weeks.

"Suppose something didn't work in the cell culture and you don't have useful information from the amniocentesis. If you do another tap and then wait another three weeks for those cells to grow, you're at twenty-two weeks. That's really cutting it close— too close. She could end up losing the option of terminating the pregnancy. Of course, we're always under some kind of pressure like that with one case or another. You just have to learn to live with it."

The blood was drawn from Sandy and her mother on the afternoon of July 19 and sent to Loughman's laboratory. The next morning, a technician prepared a culture flask for each sample and left them in the incubator until the following Thursday afternoon, when the growth of cells was harvested. Friday, the cells were killed and fixed on microscope slides. The following Monday, July 26, they were readied for shipment to San Diego. Slides from Suzanne and Stephanie prepared in the winter were included. Epstein supervised the packing and gave each slide a code number. Uta Francke would read them without any hint about which slides were from the retarded sisters and which were from Cecelia and Sandy.

Tuesday morning a delivery truck from Federal Express picked up the package and carried it to San Francisco International Airport. Later that day, one of the air cargo company's purple executive jets flew the package to San Diego's Lindbergh Field. The next day, Wednesday, the package arrived in Uta Francke's laboratory at La Jolla, a northern suburb of San Diego. She quickly treated the four coded slides with trypsin and then Giemsa stain. The next day, Thursday, July 29, she studied them under a microscope. It took her only a few minutes. She went to the telephone and called Charlie Epstein. It had been ten days.

Francke caught Epstein in his office, not always an easy task. "I have some information for you," she said.

Epstein rummaged on his cluttered desk looking for the key to the coded slide numbers. "Okay. What do you have?" he said, locating the piece of paper. He felt a tingle of apprehension. If Uta Francke couldn't do it, they would have exhausted their resources.

"SF-204 is a normal female with an apparently balanced translocation consisting of a transfer of region 5q34 to the long arm of chromosome 16," she said.

Epstein drew in his breath. The slide was Cecelia Muller's. So the extra material was coming from chromosome 5? "Okay. Go on."

"SF-205 is a female with faintly staining material translocated to the long arm of chromosome 16. She's unbalanced because all the other chromosomes are normal. She is affected, and I think the extra material on chromosome 16 was inherited from a carrier with a balanced translocation at 5 and 16."

Epstein felt a rush of excitement building. SF-205 was Suzanne.

"And SF-206 is an unbalanced translocation just like 205. SF-207 is a balanced carrier, identical to the first one, 204."

"Fantastic, Uta! Terrific!" Epstein burst out. "That's incredible. You've done it. The first one, 204, is the mother. Then 205 and 206 are the retarded daughters and 207 is the normal daughter." His words tumbled out, bumping into one another. "How did you do it?"

"Trypsin-Giemsa staining," she answered. "It's really quite clear. You'll see when I send the pictures."

Epstein felt elated. Now they could tell a balanced, unaffected carrier from an unbalanced, retarded individual. He hung up and immediately dialed Mickey Golbus' office. He relayed the news and they spent a moment congratulating one another. Golbus immediately called Judy Derstine, the genetics counselor who handled all the liaison with patients at the Santa Rosa clinic. She had become personally close to Sandy over the months and had been particularly upset at the outcome of the Monday meeting when it was decided that Sandy wouldn't receive an amniocentesis.

"I want her in here Monday for an amniocentesis," Golbus said, studying Sandy's file open before him. There was no time to spare.

"There was quite a moment of elation when Uta's interpretation of the slides matched the code numbers perfectly," Epstein recalled later. "All of a sudden the problem had opened up. We were now in a position to do something. Uta Francke had found a way to tell a balanced translocation from an unbalanced one. An amniocentesis would give Sandy a definitive answer.

"Everybody felt terrible when we had to send Sandy away without performing an amniocentesis, especially since the group had worked with her for such a long time. It's a terrible feeling to get into that kind of bind. But then to be able to recapture the situation . . . that was terrific."

The following Monday, Sandy, her mother, and Joe drove to San Francisco. Sandy was filled with apprehension again. This time there would be no reprieve from the long needle.

Joe stood by her side in the treatment room and her mother waited outside in the hall. Mickey Golbus used the ultrasound machine to determine the baby's location in the uterus and then gave Sandy an injection of a local anesthetic to deaden the skin near her belly button. He inserted the amniotic fluid needle, drew normal-appearing fluid on the first try, withdrew the needle, and placed a bandage over the insertion point. In less than half an hour it was over, and Cecelia and her daughter and son-in-law were soon on their way home. As they crossed the Golden Gate Bridge and headed north up Highway 101, Sandy began to realize how difficult the waiting during the coming three weeks was going to be.

"I beat up a girl scout once because she was saying things about my sisters," Sandy told me. "We lived in San Francisco then, at the bottom of a hill, and we had some friends around the corner. One day my sisters and I were sitting on the porch of our friends' house when this girl in a Girl Scout uniform came by. She said something like, 'Nah, nah, nah. M.R. Look at the M.R.' You know, 'M.R.' for 'mentally retarded.' I told her to stop and she wouldn't. I got mad and I began hitting her. There was a fight and I got into a lot of trouble with my parents. I was in about the second grade then.

"I've always stuck up for my sisters. They're special. They're neat. I don't think anyone has the right to knock them. They never asked to come into the world like that.

"Growing up with Suzanne and Stephie, you don't think that living with them is that bad. You're used to them. They're educable. So when I got married, at first I didn't think too much about it. I sort of felt that if this is what God's going to give me, it's what he's going to give me.

"I didn't plan on getting pregnant when I did. We had expected to wait awhile longer. But when I did, I realized it was sort of spooky, knowing there's something in the family and you don't know if it's going to happen to you. I was scared.

"Joe was a big factor in my attitudes, I think. He was an only child and his father died when he was very young. He viewed Suzanne and Stephie differently than me because he hadn't grown up with them. He was really worried about having a retarded child.

"In high school I used to go out with a friend of Joe's. And Joe remembers his friend making jokes about my sisters. When Joe was little he used to see kids make fun of my sisters. And he remembered that, too. Once before we were married we were driving around in his pickup on a rainy night. I mentioned that if we were to get married that chance was there. And he didn't say much. But I knew it worried him.

"So we decided that if the amniocentesis results were bad, I'd have an abortion. It seemed hard enough bringing up a normal child, but to have an extra thing like that. But it was hard waiting during those weeks. Finally, when it had just about been three weeks, I called Judy Derstine. She said she hadn't heard anything but that she would check and call me back. She called back to say it would be just a couple more days."

The sample of Sandy's amniotic fluid containing cells sloughed off by the growing fetus was sent to Bill Loughman's laboratory, to Gisella Halbasch, a specialist in growing amniotic fluid cells. Golbus performs amniocenteses each Monday and Thursday afternoon, and Gisella spends the following late afternoons and evenings initiating the cell cultures. She divides each fluid sample and begins two cultures for each patient. The fluid is placed in flasks and a mix of nutrients added that long experience has shown will promote establishment of thriving cell lines. The nutrients in the flasks are replenished and the cultures given careful attention while cells multiply until there is a sufficient number for study.

The cell colonies are harvested when Gisella considers their growth sufficient, and then sent through the same fixing and staining treatment that the blood cells are given to read the chromosomes. The cells from Sandy's baby were given special attention, including a trypsin-Giemsa stain.

Amniocentesis results are analyzed by Golbus, Loughman, and the technicians each Friday afternoon. A flurry of telephone calls follows as the genetics counselors relay the news to the patients. As is the case in all genetics centers, most of the news is good. Of the twenty or so patients who must be telephoned each Friday, no more than one or two in a typical week receive bad news. It is Golbus' policy to telephone these patients himself. In earlier years, before the UCSF genetics group's load grew to over one thousand amniotic fluid taps a year, Golbus did all the telephoning.

Judy Derstine normally wasn't involved in telephoning amniocentesis results. But she had asked to be permitted to handle Sandy's. Through the afternoon she tried to reach Sandy without success. The genetics counselors take home the files of the patients they haven't been able to reach by the end of the day and keep trying through the weekend. Judy took Sandy's file to her apartment in Mill Valley across the Golden Gate Bridge in Marin County. When she telephoned about 7 P.M., Joe answered on the wall telephone in the kitchen. Judy asked to speak to Sandy.

"It's Judy at the hospital," Joe said, handing the telephone to his wife.

Sandy could feel her heart pounding. Her stomach knotted up. She took the telephone. "Oh! It is?" she shrieked after listening a few seconds.

"Joe!" she shouted at her husband, who stood in the kitchen door looking at her. "It's normal. It's normal. There's absolutely nothing wrong with it."

"Do you want to know the baby's sex?" Judy asked. A chromosome study always reveals the sex, but not all patients automatically want to learn what it is.

Sandy thought a minute. "No, I don't," she said. "I want it to be a surprise."

After she hung up, Sandy and Joe held one another for several minutes. Sandy's cheeks grew wet with tears.

It was a foggy day the following January when Judy Derstine hung up the telephone in her office, walked down the long hallway from her building into Moffitt Hospital, and took the elevator to the pediatrics department on the sixth floor. She was looking for Charlie Epstein. His office door was open, but he wasn't in sight. The secretaries across the hall hadn't seen him. She returned to his office and picked up a note pad. "Charlie," she wrote. "Sandy had her baby girl Jan. 7—8 lbs., 11 oz. & everything OK! Judy."

THE SINGING
HALLS

The first yellow leaves of the South Carolina autumn swirl across the highway in a gust of wind, dancing before the headlights of passing cars. The shafts of light momentarily sweep over the portable sign with clip-on letters standing by the side of the road in front of the Bennettsville High School auditorium. It proclaims:

GOSPEL SINGING TONIGHT

7:30

EVERYONE COME

Inside the auditorium, two hundred rough-hewn farmers, storekeepers, factory workers, and widows and old people on Social Security sit on the hard wooden seats, their attention focused on a large man in a vested black pinstripe suit. He stands on the small stage amid the paraphernalia of traveling musicians —bulky amplifiers, giant loudspeakers, drums, steel guitar, and microphone stands. A local theatrical group plans the following weekend to perform *The King and I*. The stage's backdrop is a crude representation of the hall where Anna and the King of Siam dance. Purple gels filter the lights overhead, giving the stage a darkened, intimate feeling.

The big man is speaking, the microphone held close to his mouth. The white spotlight in the back of the auditorium, operated by obviously inexperienced hands, casts a narrow circle of light on the man's face. He has graying hair combed forward to hide a receding hairline, a small graying mustache, and a paunch over which the pants and the vest of the suit are stretched tight. His voice is rich with his southern heritage, a mixture of Jim Nabors and Andy Griffith and Jimmy Carter. It is a voice that he can control perfectly to command or cajole, to plead or exhort. He is a master at using it to wring just the emotion from a crowd that he wants, to play the people as if they were a piano, knowing exactly which keys to strike to produce the chord he desires. He is a born-again Christian and he uses his voice in the service of his Lord and Master Jesus Christ. His name is Jerry Hall, father, husband, promoter, and guiding light of The Singing Halls, gospel singers who promise every crowd "more than entertainment." He is introducing his family.

"I want you to meet our oldest son, Steven," he says in his rich drawl. Steven rises from his seat in front of the piano. "Steven is eighteen and a delight to his parents. A fine, born-again Christian."

An "Amen" or two drift up from the audience.

Jerry lowers his voice and shifts his body forward, adopting a conspiratorial tone. "And I want to tell you girls out there something," he says. Pause. "Steven is *single*." Laughter ripples through the room. His voice suddenly booms out, commanding. "But y'all better hurry, 'cause there's a whole passel of girls already achasin' him." The laughter grows. People shift in their seats, casting smiles of pleasure at their neighbors. This is what they've come for—entertainment, the Gospel, and the message that Brother Hall and his family will bring.

"Now this here fellow is gettin' ready to pick his knee," Jerry is saying, pointing to a young man who stands in front of a microphone, an electric guitar on a long strap hanging just above his knees. "This is our youngest son, Stanley." Stanley bows. "Stanley is eighteen and also a fine born-again Christian. He is ten minutes younger than Steve. He's the baby of the family." The crowd titters. Stanley is bigger than Steve, and the guitar and his long arms give him a hulking appearance.

Jerry's voice becomes sad. "I'm sorry to tell you young ladies that Stan is spoke for." Laughter. "That's right. This young'un is spoke for. This little girl come pesterin' me and Sara, crying to us and carryin' on so you wouldn't believe. She wanted us to *give*

Stanley to her. Let me tell you, she just wouldn't give us no peace. So finally we said, 'Oh, here. Go ahead and take 'im.' So, girls, Stanley is engaged. They'll be married next June."

He steps forward and holds a hand to his brow, as if looking past the lights, scanning the auditorium. "Where is that girl," his voice booms. "Pamela! Stand up out there and let these folks take a good look at you!"

The spotlight sweeps wildly across the crowd. People strain forward in their seats, sensing movement in the aisle toward the front of the room. A woman has risen from a seat and stepped into the aisle. The spotlight sweeps past her, returns, misses its mark, jerks back, and finally stops to reveal a tall, willowy young woman with brown hair cascading below her shoulders. She is wearing a long dress of billowy material. A paper rose is in her hair. A few hours earlier, when she modeled her new dress for Jerry, he gave Stan a five-dollar bill and sent him into Bennettsville to buy a black scarf to make the dress's bodice more modest. She waves.

"There she is, folks. That's the girl that wouldn't give us no peace till we give Stanley to her. That's our future daughter-in-law, a fine Christian girl, and we love her very much."

He quickly introduces Jerry Lloyd, the group's steel guitar player, and H. W. Corneal, the drummer. "H.W. is from Richmond," Jerry says. "He's with the POE-LEECE. If yer goin' to Richmond, he's the one that's gonna git ya. In fact, if yer goin' to Richmond, he's done got ya."

Finally, he turns to the plump woman in a flowing pink gown who stands beside him. He puts his arm around her. "And this here little lady I've been introducin' nearly thirty years. This is Sara." He pronounces it *Say-rah*. Sara gazes into his face.

"Say-rah come and took me from my momma when I was just a little boy. She come pesterin' my momma, wantin' her to go and give me to her till my momma just couldn't say no. My momma didn't know what to do with me nohow." He pauses. The crowd is chuckling. "A few days later, Say-rah didn't know what to do with me neither." Laughter. Another pause, perfectly timed. "But by then, my momma wouldn't take me back." The crowd roars.

"We have someone else we want to introduce you to," Jerry

adds. "His name is William Stockton. He's from Copake, *New York*. After the service tonight, I want all you folks to go out and buy sympathy cards and send them to this boy for havin' to live in a place like that. It's not his fault. Maybe he doesn't know no better." The crowd is chuckling.

"Where is that boy?" he calls, peering into the crowd. "Stand up out there, son." The spotlight is sweeping across the darkened auditorium again.

I stand up and the light settles on me. I raise a hand in an embarrassed greeting.

"Now there, folks, is an honest-to-goodness Yankee. Take a good look. When was the last time you saw a real Yankee in these parts?" People are turning in the darkness, rising in their seats, straining to get a look. "He's a writer and he's travelin' with us this weekend to write about us. I want you folks to be real nice to him, even if he is a Yankee boy. And you kin sit down now, Bill."

An old man in the row in front of me turns and offers his hand. A woman behind leans forward and pats my shoulder. "Nice to have you here, son," she murmurs.

Now Jerry moves forward to the edge of the stage, leaning forward, taking the crowd into his confidence by the way he holds his body. This is what they've been waiting for. People lean forward expectantly. They have heard of this man and his family and his story.

"Those of you who know our family or know about us, know that someone is missing tonight." His voice is sad now. It almost trembles. The crowd suddenly becomes so hushed that a small baby's babbling rings through the room. "Our daughter Susan is missing. Last July 24 she went to be with her Lord. We know with a conviction born of our faith that she's there, that she's happy." He steps closer to the edge of the stage. "Amen," Sara calls.

Jerry's voice is soft, brimming with tears. "Oh, Lord, how we miss that girl so. How empty our hearts are that she is gone. Those of you who were ever privileged enough to hear that girl sing, to experience her witnessing to the power of our Lord Jesus Christ, the Son of God, know how much our hearts cry out to her. She's with her Lord in Gloryland." His voice is hoarse. Suddenly it booms out. He raises his arm. "But oh, it makes us happy to know that she's there walking those streets of gold waiting for us. How Sara and I and the boys look forward to that day when we can join her, when we can be reunited."

Chills are running up and down my spine. I have never experienced anything like this. I can hear women behind me sobbing. A man two rows down is blowing his nose. The sound of sniffling rises up all around the room. Sara is dabbing at her eyes with a lace handkerchief.

Jerry's voice is soft again. "The boys miss Susan so much, too. They've taken an old song you all know and rewritten it just for Susan. Here it is."

Steve's fingers caress the piano keys. Jerry Lloyd makes the steel guitar wail softly. H.W. begins a beat on the snare drums. Stan strums the electric guitar. Steve bends over the piano, his mouth close to the microphone positioned above the keys. His rich baritone fills the room, the others joining in.

> *Susan's teaching angels how to sing,*
> *Heaven draws a silence when Susan sings.*
>
> *David lays down his ark*
> *And heaven's bells don't ring.*

The tension in the song swells steadily, the chorus ringing after each verse. In the spotlight, tears are streaming down Jerry's face. Steve is bent over his microphone, his eyes shut tight as he plays and sings. Sara repeatedly wipes her eyes.

People begin to come to their feet as the family sings. Some are openly sobbing. Others furtively run the side of a hand along their cheeks, wiping back the tears.

The Singing Halls have begun another performance of their gospel music.

"We've spent fifty months sittin' in hospitals waitin' for one of our children to die, honey," Sara says. "Four children, and they've all died of leukemia. But I don't look at dying like most folks do. Dyin' is just part of livin'. When someone is gone, all you have left are the memories. If the memories are good, then you'll be all right. If the memories are bad, you're gonna have problems. You know what I mean?"

It's nearly midnight and the old Greyhound bus, purchased by the family and refurbished with bedrooms, a refrigerator, and a living room, thunders down Interstate 95 toward Charleston, South Carolina. They'll drive all night, taking turns at the wheel.

We sit in the living room, directly behind the driver's seat. It consists of a couch along each wall with an aisle between. The bus is richly furnished with shag carpeting on the floors, walls, and ceiling. Overhead, built into the storage space where travelers once threw their suitcases, are a television and tape players and a clock whose face is one of the family's LP phonograph records.

Jerry is reclining on the couch next to his wife. Steve is driving. Stanley and the others are in the back sleeping. Pamela has left the group, returning to her home near Charlotte, North Carolina. She travels with the family at every opportunity. It is one of Jerry's conditions of the impending marriage. She must experience their life and convince herself and her future father-in-law that she can fit in.

"I was nineteen and Dad was . . ." Sara frowns and looks over at Jerry. "Dad, how old were you when Sammy got sick?"

"I was just a little boy," he answers, grinning.

"Sugar, be serious now. This man has asked a question. I was nineteen and he's five years older. So ya must've been twenty-four," she concludes. "That was in 1955. We were babies. We didn't know anything. We was livin' in Florida then. I can still remember that old doctor. He was an old man, probably in his seventies. He had white hair and looked like a doctor should look. You know what I mean? He said, 'Children.' He always called us 'children.' And we weren't nothin' but children then. He says, 'Children, lightnin' don't strike twice in the same place. You'll never have to go through this again.'

"Then when Scott got sick in 1962 they said, 'Yes, there have been families that's had two cases of leukemia. But ya can rest easy now that it won't happen again. Then Sandy got sick in 1968 and then Sue in 1976. Each of my children has got sick and died about seven years apart."

Jerry raises up on an elbow. "It wasn't till we got to the National Institutes of Health in Maryland with Sandy that they told us it might be somethin' in the family," he says.

If they had known when they married what they know now, I ask, would they have altered their plans about having children?

"I sure would," Jerry says emphatically, sitting up. "If I knowed then what I know now, I wouldn't have children. I wouldn't. All the sufferin' they've been through . . . we've been through."

"Oh no. No," Sara says. "No, Dad. How can you say that?" She turns to me. "How can you say that, Bill? Even if they could tell you that might happen, how could you say you wouldn't have children?" She turns to Jerry. "Sugar, think of all the love they give us, all the joy. And now they're with their Lord. Praise the Lord."

"If they'd found out earlier, we couldn't 've done nothin', anyway," Jerry adds. "By the time Sammy was sick Scott and Susan had been born. And by the time Scott was sick Sandy and the twins had been born."

We drive on in silence, the bus's headlights picking up the green interstate highway signs. We're alone on the road with the big eighteen-wheeler trucks, which steadily pass us in the left lane.

Sara breaks the silence. "I had triplets. Three little boys. They would be fifteen now. They were born at eight months and they all died. Maybe it was better that way. The Lord knows what he's doin'. It would've just been three more to worry over.

"A few months after that, they found I had cancer of the cervix. I was twenty-seven then. They gave me radiation. Lord, how that made me sick. Oh, honey, you can't imagine it. Made all my hair fall out."

Jerry speaks. "About four years ago Sara went for a Pap smear, and when the results came back the doctors saw some cancer cells. We thought Sara's cancer had come back. She made an appointment to go to the university at Chapel Hill. Between the time she had the first test and goin' to the university, every congregation we sang to I told them about Sara's cancer and asked 'em to pray for her. When she went to Chapel Hill they did another Pap test and when that came back there wasn't no cancer." He fixes me with his gaze. "What do you think of that?"

I don't answer.

"God did that. That's what."

Looking at Sara Hall, you can see the beauty that attracted Jerry when Sara was just fifteen, the daughter of a poor North Carolina tobacco farmer with eight children. They ran away to be married and then lived at home with their respective families for six months, keeping their marriage a secret.

Sara and Jerry have had an emotionally turbulent life. Sara had cancer, but survived it. Four of Sara's children have died of acute myelogenous leukemia, a cancer of the bone marrow. Her father, an aunt, the aunt's son, an uncle, and two children of one of her sisters all have died of cancer. In 1977 another sister was diagnosed with breast cancer. It is not surprising, then, that she looks tired, weary from a burden pressing her down. It is plain that their devout faith in God and Jesus Christ is what "keeps us keepin' on."

Everyone who meets The Singing Halls and spends any time with them comes away impressed. They are an extraordinary family. Some of the doctors at the National Cancer Institute in Bethesda, Maryland, feel the same way.

"Jerry Hall is a man of great inner resources," says John Mulvihill of the National Cancer Institute, one of the institutes within the National Institutes of Health. He is a doctor trained as a pediatrician and geneticist who brings together genetics, cancer, and epidemiology, which is the study of how patterns of disease outbreaks are related to causes of disease. He had considerable contact with the Halls during the two years Susan was dying of leukemia. Like many of the NIH physicians, he found the Halls so ingenuous that he had to restrain himself and make certain he kept a professional distance from them lest it interfere with his caring for them as a physician.

"In general, we don't know what causes cancer," says Mulvihill. "There is a lot of publicity now about things in the environment that cause cancer. But everyone's experience is that their Uncle Joe smoked three packs of cigarettes a day all his life and died at age eighty of a heart attack. So we know there is some difference from one person to the next in terms of getting cancer.

In some rare cancers, genetic factors are very conspicuous. Retinoblastoma, which is a cancer in the retina of the eye in children, is an example. In other cancers there is a vague tendency for that cancer to run in families. Breast cancer is one. If one woman in a family gets breast cancer it puts the other women in the family at a three times greater risk for breast cancer. Even the commonest cancers—breast, lung, colon, prostate, and stomach—show some familial aggregation.

"A problem that faces geneticists in many areas and not just with cancer is distinguishing the genetic factors from the environmental factors that cause the disease. A classic method to look at this question has been through twins. Identical twins are identical genetically. Fraternal twins are not genetically identical, but, in general, share only one half of their genes. Therefore, if you study how often a disease occurs in both members of sets of identical and fraternal twins, you get some idea about genetic factors and environmental factors.

"There have been two groups of twins, one in the U.S. and one in Denmark, that have been looked at in relation to cancer. But there was no striking relationship. The problem is that individual cancers have been too rare among the twins to answer the question. Some relationship may emerge as the twins get older.

"In the case of the Halls, it's either the environment or their genes or the interaction of both. The family has been very mobile over the years. Different branches of the family that have developed cancer have lived apart, with little contact. After questioning them about known things in their surroundings, we haven't been able to identify any environmental agent that was unusual. So the indication is that their cancers are genetically caused, not environmentally caused.

"Is there anything in their genes? That would be answered by either finding an inborn error in their metabolism or finding a chromosomal abnormality or a cellular defect. Some of the research, which took place before I came to NIH, involved looking for differences in their cells, and there were some intriguing results.

"One experiment conducted in 1969 at the time Sandy was ill involved taking skin cells, or fibroblasts, from each family member and growing them in tissue culture so that you had a line of cells. Then the cells were treated with SV40 virus, which is a virus found in certain species of monkeys that causes cancer in experimental animals. The SV40 virus can also cause cells in a tissue culture to become malignant.

"So the cells from Jerry, Sara, Sandy, Susan, and the twins were treated with the virus and the rate of transformation of normal cells to cancer cells measured.

"The experiments showed that Sara and the two girls' cells were more easily transformed into cancer cells, from five to fifty times greater transformation. The cells of Jerry and Stan and Steve were transformed at a normal rate.

"Those results were intriguing, especially because the virus transformation was in a connective tissue cell—that is, a skin cell —while the cancer that affected the girls was in a different kind of cell—a bone marrow cell. But the SV40 test was what we call a bioassay and many bioassays are rather capricious. Repeated studies since then of the Halls' cells using the SV40 virus assay haven't held up.

"Dr. George Todaro of NCI did the original work. But he has other, bigger cancer virology efforts under way and he hasn't been able to pursue it further. Other less famous laboratories have taken on the job and gotten equivocal results. So no one has actually begun from round one again and said, 'Let's do it just like Todaro did and try again.' Meanwhile, it's entirely possible the SV40 virus used originally may have changed and won't work anyway.

"But we have cell cultures from Susan and Sandra in storage, along with Susan's urine and blood samples. We also have samples from other family members in storage and could go back to them for additional samples if necessary. We hope to keep track of new developments in leukemia and cell biology and see what new research could be applied to these cells.

"One new method was developed about two years ago and wasn't even available when Sandra was alive. It involves measuring how much radiation damages a cell in tissue culture. This, in turn, reflects to what extent there might be errors in the biochemical machinery inside the cell's nucleus that repairs defects the radiation causes. There are suggestions that errors created by this repair system in a cell could be related to the development of a cancer cell. So we contacted a laboratory in Canada which has established an international reputation in this assay and asked them to help us. They did find in both the girls and the mother that their fibroblasts were somewhat more sensitive to radiation and presumably more prone to repair errors than normal fibroblasts. The twins and their father were normal. The Canadian lab is taking it further to see if a repair defect is actually involved.

"We also sent their urines to Tokyo to Japan's National Cancer Institute, where they are performing studies with bacteria that are especially easy to change genetically—that is, mutate—by treating them with certain chemicals. The assay is being touted as a quick first screen to see what chemicals may be carcinogenic.

"So the question is: are the Halls putting out anything in their urine and presumably in their blood that could cause mutations within cells and is exposing their bone marrow cells to frequent transformation to cancer? This assay is positive, for example, in cigarette smokers. Somehow, smokers get enough of the metabolites of tobacco into their blood and urine to cause cells in culture to mutate and make this test positive.

"We sent the Japanese scientists some specimens that are coded with numbers so they don't know which samples are from a family member with leukemia and which are from Jerry and Sara and the boys. That way, there's nothing to bias their results. The Tokyo lab sent word recently that they are seeing differences between the samples. There is still much more of this research to be done in their lab and we don't want to break the code yet. But it's very exciting waiting to learn if they have a new finding or not. However, we're being cautious about what we might learn from the Japanese tests. It's a bioassay and we've been burned by bioassays in the past, so we want to keep these experiments as rigorous as possible.

"We also have sent specimens from the Halls to several other labs. We want to hold an all-day meeting sometime in 1979 to get all these different researchers together. We can have sort of a brainstorming session to say, 'Well, we didn't think these results were significant, but in view of your results, perhaps we should do the following.' Maybe together we can all come up with something new.

"At all of these labs this work isn't their primary research interest. They're squeezing it in. But we feel that their assays should be attempted. Most of the scientists involved see the merit of these fishing expeditions. We're firm believers in the serendipity of science."

Sandra, the Halls' third child to contract acute myelogenous leukemia, was ten years old when doctors in North Carolina diagnosed her disease in 1968. She was referred to the National Institutes of Health's clinical center in Bethesda, Maryland, where there is a ward set aside for leukemia patients.

The NCI doctors were interested in the family's history of cancer, particularly when detailed questioning revealed so many members of Sara's family with cancer. An obvious research avenue was to look for biological differences between Sandra, who had the disease, and her father and mother and three siblings, who did not.

One question the geneticists sought to answer was whether the cancer was being passed from generation to generation in a dominant or recessive fashion. The answer to that question depended upon whether the inherited disease was considered to be cancer in general or just acute myelogenous leukemia, which doctors often call AML for short.

With one exception at the time of Sandra's illness, every cancer case in the family's experience had been on Sara's side. The exception was Jerry's older brother, S.D., who died at age seven of bladder cancer many years before Jerry was born. On Sara's side, at the time of Sandra's illness, there had been several cancer deaths, including some cases of AML.

Sara's father, Rufus, died in 1956 of cancer of the kidney. Her great-aunt on her mother's side, Sally, died in 1957 of AML. Sally's son William died of AML in 1961. A year after Sandra's disease was diagnosed, Sara's uncle Billie, brother of Sally, was diagnosed with AML. He subsequently died. Two of Sara's nephews, sons of her sister Mary Louise, died of a cancer called malignant reticuloendotheliosis, a cancer related to the family of leukemias. The first nephew, Howard, died in 1952; the other,

Steven, in 1964. Thus, when Sandra died of AML in 1970, four days past her thirteenth birthday, three of Sara's children and three close relatives had succumbed to AML. Three others had died of other cancers. There was no AML on Jerry's side of the family.

If the doctors considered the genetic trait being passed on to be cancer in general, then it seemed to be dominantly inherited, since virtually every case was on Sara's side of the family. A dominantly inherited disorder can pass down one side of the family without the spouse being involved.

But half of Jerry and Sara's children had contracted AML, a percentage that also agreed roughly with what might be expected of a recessive disorder, in which each spouse must contribute some of the genes involved to produce, not just any cancer, but only AML. But what gene or genes might be involved or whether the disease was a recessive or dominant disorder was far from clear.

When the experiments with Hall family members' cells growing in tissue culture revealed a difference in response to the cancer-causing monkey virus, there was an obvious temptation to use the results to predict whether other family members would develop leukemia or some other cancer. Some of the NIH doctors were reluctant to do this. However, physicians must tell a patient something, particularly when the patient is a man like Jerry Hall, who badgered the NIH doctors unmercifully for as much information as possible. So the Halls were told that a case, albeit a flimsy one, could be made for the possibility that Susan was at risk for AML, but that Steve and Stan weren't. That was in 1970, a few months before Sandra died. At the time, Susan was fourteen, a beautiful, charming young girl. Eight years later, she, too, would die of AML.

"Yeah, they told us Sue could get it. They told us all right," Jerry says. He is sitting on the edge of the couch across the aisle from me, his feet planted, knees spread wide, a palm on each knee. "I didn't believe 'em."

He fixes me with his direct gaze. There's just enough light in the darkened bus so that I can see his eyes. It's difficult to look away. "I didn't believe 'em because they couldn't tell me that

Sara wouldn't get AML either. Sara's cells reacted to the SV40 virus just like Sue's and Sandy's. So if Sandy had AML and Sue was agonna get it, how did they know Sara wouldn't?"

"Dad?" Steve's voice comes from the driver's seat. "Do I take this exit here?"

"You're drivin' the bus, son," he calls, not wanting to be distracted.

"Dad! I don't know this route," Steve answers irritably. It is the first time all weekend I've heard a sharp note in his voice. In fact, it's the first sharp word I've heard between family members. The boys always address their father as "sir" and are as deferentially respectful as a private in the presence of a general.

"My goodness, boy," Jerry says, rising. I wonder if he is about to rebuke his son. Instead, he moves forward, clasps Steve fondly on the shoulder, and bends down to see through the windshield. "Yes. That's the exit. Twenty-six. Take it. It'll shoot us straight into Charleston. You're doin' fine, boy."

The green exit sign is looming up at us and Steve begins frantically shifting to lower gears and pumping the brake to slow down so the bus can take the exit ramp. Jerry lurches and almost loses his balance before returning to his seat.

"Those doctors at NIH are all nice folks—well, almost all," he continues. "And I know they mean well. They really do. But, Lord, you can't get a straight answer from any of them. 'Well, Mr. Hall,' they'll say. 'It might be this or it might be that. But on the other hand, it could be this.' Lord have mercy, you don't know what to believe."

We drive on in silence. The new road is another straight, four-lane highway. The headlights hint at the forests of jack pine and swamps filled with dank water that stretch away into the distance.

"When did you first know that Sue had AML?" I ask.

Sara has moved and is reclining on the couch beside me, her legs wrapped in a blanket. Jerry and the boys always keep the bus too cold for her.

"Bill," she says, sitting up and touching my arm. She pronounces the name "Beel." "Honey, I know what caused Sue's leukemia. Yes, I do," she says emphatically, responding to my skeptical look. "It was a tetanus shot. She went and got a tetanus shot, and six months later she had leukemia. She knew better than to do that. She shouldn't 've done it."

It sounds like a crackpot notion. Seeing my look, Jerry explains.

"Sandy and Sara and then Sue were all allergic to medicine. They get terrible reactions to havin' medicines. You know, injections of antibiotics. Sandy and Sue had bad reactions to many of the cancer drugs they give 'em at NIH. Worse reactions than most other folks on the ward. When Sandy was dyin' a doctor at NIH told us, he said, 'I'm not sure about this and if you ask me tomorrow I may say I never said it, but if I was you I wouldn't allow no one to give anybody in your family an injection of anything.'"

"You mean because it might have an effect on the immune system?" I ask.

Allergies are related to reactions of the body's natural defense system—the immune system—to some foreign substances introduced into the body. In addition, there is a growing body of scientific research that relates the immune system to cancer. Immune system cells that ordinarily seek out an invading organism and attack it often ignore cancer cells, even though they should be recognized as "foreign" and attacked. When they studied the effect that the monkey virus had on the Halls' skin cells, the doctors also looked for differences in their immune systems, but found nothing conclusive.

"You see, sugar, Sue was a grown woman and married and living in Sanford, North Carolina, before she got leukemia. Dad and I and the boys were in Charleston," Sara resumes.

"She married the son of a doctor Dad and I've known all our lives. He delivered a couple of my babies. So she was grown and we didn't have no control over her. One day they was movin' a picnic table out of the house and she stepped on something. So she went to the hospital and had a tetanus shot.

"It was about six months later that she was in a store in Sanford trying on a dress. When she raised her arms up to put on the dress, somethin' happened to her back. She was in terrible pain. They said she slipped a disk. They put her in the hospital in traction. Now, we was in Charleston and I couldn't just run up and see her. But she told me on the phone that she was so tired.

"She went home from the hospital and one day she called again and said how tired she was. She'd been out to the clothesline and it was all she could do to go out there and back. She said, 'Mama, somethin's wrong with me.' So I said to Jerry, I said, 'Dad, somethin's wrong up there. I'm goin' up there.'

"I knew she had leukemia when I took her in my arms to hug her. I could smell it in her hair. It was the same smell that was on Sammy and Scott and Sandy. Later, after we was at NIH, they sent around someone who was expert in smells and she had all kinds of little bottles of things that I smelled. I said it smelled like geraniums or the smell you get sometimes at a funeral when the scents from the flowers are mixin'.

"Well, to make a long story short, I took her to the university at Chapel Hill and they done some tests and then she was in the hospital at Sanford and they made some tests. I think those doctors knew right away that she had leukemia and they deliberately didn't tell us for two weeks, but that's another story for another time. But when they did finally say it was leukemia, we called Dr. Levine at NIH, Dr. Arthur Levine, the head of the leukemia ward, and he said to bring her right up.

"Sue was mighty sick by then because those doctors at Chapel Hill and Sanford wasted two weeks. But Sue was so tough, Bill. She went up to Bethesda and in just a few weeks she walked out of there in remission. Her husband went up to Bethesda for the first two weeks."

Jerry interrupts. "He spent two weeks there. One Friday he said, 'See you Monday.' And came home. We're still waitin' for Monday."

"He walked out on her is what he did," Sara says. "Sue took it real well although it liked ta killed her at first. I don't know why, but it became real important for her to get a divorce. She wanted her name back. She wanted to be a Hall agin. She studied on that real hard. In April before she died she got her divorce. She was happy then. She was Sue Hall agin."

Jerry takes up the story. "You see, the problem was that Sue was a grown woman. I didn't have control over her. But I couldn't oppose her. All you have when they're gone is your memories. I wanted those memories to be good uns.

"She was in and out of that hospital in Maryland several

times. There was times when the doctors didn't think she would live another two weeks, and then two weeks later she would be well enough to walk out of there. There was a period when we come back down to North Carolina, and when it was time to go back to Maryland, she said she wasn't goin'. We were in the car drivin' north and she said she wasn't goin'. Said to turn around and go back home. If we didn't she would open the door and jump out. And she would've, too. Wouldn't she've, Mother?"

Sara nods her head.

"I turned the car around and went back home. I wanted those good memories."

"Oh, Dad, you always petted Sue. All her life," Sara interrupts. "He did, Bill. He did," she adds, turning to me.

"We come back home and Sue wouldn't take her medicine, she wouldn't do anything. She got real sick," Jerry continues. "We just didn't know what to do with her. One night I said to Sara, 'Tomorrow, I don't care what she says, I'm takin' her to Maryland.' We got down on our knees and prayed to God, askin' him to help Sue and to help us. And you know what? The next mornin' Sue says, 'Daddy, I want to go back to the hospital.' Now, what do you think of that?

"She was real sick by then. I called Dr. Levine and he said, 'Bring her on.' I told him how sick she was and he said somethin' on the phone kind of made me mad. He said, 'Are you prepared for anything that might happen?' I said, 'What d'ya mean?' He meant was I prepared in case she died in the car on the way. It made me mad because I thought he didn't think I had enough sense to know what to do.

"I said, 'Sure I'm prepared.' And he said, 'What are ya gonna do if she dies on you. Turn around and go back? Stop and find a hospital? Or bring her on to here?' I said, 'I'll bring her on to you.' He said, 'Okay. We'll get ready for that possibility, too.'

"We drove up Interstate 95 in the car goin' ninety miles an hour. The Lord was with us. We only saw but one policeman and that was in Virginia goin' the other way. I got her to Bethesda inside of four hours. They was waitin' for us. Had everything ready. Ya know, weren't but a few weeks and Sue walked out of there agin in remission."

I asked John Mulvihill, the cancer epidemiologist, about the odor in Sue's hair.

"One of the beauties of being at NIH is that there is some scientist tucked away who is into everything," he answers. "When we heard about the odor, we were able to find a doctor who specializes in smell chemistry. We described the odor to her and then she talked to Mrs. Hall. They identified it as geraniums.

"About a week later, this doctor, whose name is Helen Lloyd, saw Sara in clinic and brought down a bunch of bottles and Sara said it is most like the smell in this bottle, but not exactly. So Dr. Lloyd went back to her lab to modify the chemical to see if she could get it right on the nose. She found that it was some compound in the phenylalanine-tyrosine metabolic pathway in the body. Phenylalanine and tyrosine are amino acids. We still don't know which chemical it is with certainty because Susan lost the odor soon after treatment began and it never came back.

"There has been some evidence incriminating tyrosine abnormalities with childhood cancer, mostly liver tumors. We couldn't find any evidence that AML is involved. It's the same metabolic pathway that's involved in phenylketonuria and other inborn errors of metabolism. How it relates to leukemia we don't know. We don't know how it could be involved in the transformation of a normal cell into a cancer cell except that as an amino acid it is involved in protein synthesis. It is possible a cell might express some abnormality involving this protein synthesis and that somehow would lead to a transformation to a cancer cell.

"If you consider that the Hall children are expressing a recessive genetic condition, then it's possible it could be an inborn error of metabolism leading somehow to the AML. Most inborn errors of metabolism are inherited in a recessive fashion and involve a defect in an enzyme that controls some biochemical process. On the other hand, the odor certainly could be a red herring.

"In fact, there are lots of red herrings the Halls have tossed us. For example, we can't tie in the allergies that Sandra, Susan, and Sara have. They're fairly spectacular allergies to all sorts of medicines. Many of the medicines Susan received caused terrific skin reactions and body reactions. The reactions had been seen before with those drugs in other patients, but every drug Susan received seemed to give a reaction. It impaired her treatment. Sandra had the same type reactions.

"But I've learned not to write off anything the Halls say. Certainly the doctors providing therapy to Susan learned that the Halls knew more about leukemia therapy than some of the doctors who were supposed to be taking care of Susan. In terms of etiology—finding the cause of their disease—I don't want to write off anything.

"About their conviction regarding Susan's tetanus injection, all we can say is that we can't think of a reason why that's related. Most cancers have a long latency. Based on the experience of the atomic bomb survivors in Japan, the excess of leukemia started rising about two or three years after the bomb and continued at a high level for about three or four years and then began to drop down. So it seems that where there is a radiation cause of the sickness you need at least three years.

"In the case of workers who were manufacturing the chemical benzene, they had long exposures of rather heavy doses of benzene before the leukemia that chemical causes began to appear.

"On the other hand, can we say that we've studied a thousand people who got tetanus shots and followed them to see if they developed leukemia? The answer is no. Such a study hasn't been done. All I could tell Jerry is that there is no evidence that the tetanus shot is related to Susan's getting leukemia. However, certain medicines are incriminated in aplastic anemia, which is a disease similar to leukemia. Chloromycetin, which is an antibiotic, is one. Phenylbutazone, an analgesic, is another. It's a good principle for everyone to take medicines only when they're needed. But in the end, we can't say anything definitive in relation to the Halls.

"There is some question about whether there is a relationship between the stillborn triplets and Sara's cervical cancer, which showed up several months after her miscarriage with the triplets. We don't know what that means. It's intriguing to note that the pregnancy before the stillborn triplets involved twins. There is some relationship between being a twin and being at risk for leukemia. It's mostly a statistical and epidemiological relationship. There are a fair number of cases in which one twin got leukemia and then the other did. There is also some evidence that the mothers of leukemic children seem to have more twins than women without leukemic children. It all ties in vaguely, but we have no biological interpretation of what might be occurring.

"Jerry's also told me that he didn't believe the SV40 assay results and the prediction that Susan would develop leukemia because no one would assure him that Sara wouldn't get leukemia. All you can say is that if the end point of whatever is happening in the Hall family is the occurrence of a cancer, then Sara's already been at risk and had hers. But we don't know that. So you can't say she isn't at risk for leukemia. Also, you have to remember that any woman who's had cervical cancer is at higher risk for other cancers.

"I would not base any genetic counseling for the twins on the results of the SV40 tests. You've heard Jerry and Sara say that after the first child had AML they were told it wouldn't happen again. And then the second one had it and they were told, 'Yes, there are some families that have had two cases.' And then the third child got it, and so on. So I'm not about to tell them that the boys aren't at risk.

"The boys have almost a totally clean slate for all the biological tests we've done. Nothing has been abnormal in them. One had an abnormal amino acid when one lab did a study. But then when we contracted with another lab to repeat the test for that defect they weren't able to find it. We now feel that was a lab error. Nothing in the boys' immunological profile and their genetic markers would indicate they are at risk. But we still can't counsel them genetically about it."

Sara has fallen asleep on one of the couches, a blanket pulled up over her. Jerry and I sit side by side on the other couch, each of us turned sideways with a leg tucked under so we can see one another. Steve is still driving and the big trucks still rumble past us in the left lane.

"How're we doin' on fuel?" Jerry calls.

"We have enough to get to Charleston," Steve replies.

"This old fat boy was nothin' but a drunk when Sandy got sick back in 1968," Jerry suddenly says. "Sara flew up there with Sandy and I followed a few days later in the car with Susan and the boys. Sara and I spent thirty-three days in that ward with

Sandy without ever takin' our clothes off till things settled down.

"We moved to Maryland and had an apartment so we could be with Sandy. Somebody was always with her. We did the same with Sue.

"I was drinkin' all the time when Sandy was sick. We didn't have any kind of religion, Sara and me. But Sandy did. She got it in Sunday school. I was obnoxious. I was just a real blankety-blank to everybody at NIH. I must have been somethin' else. Those doctors were all afraid of me. You know where I learned how NIH worked? Where to go and who to see about what and how to git somethin' done? You know who I learned that from? A janitor. That's who. I got to know those kind of folks real well. And they would give me advice. They'd come to me and say, 'Mr. Hall, if you want that for your daughter, all that doctor has to do is sign his name to the form. Tell him to sign his name.' Those kinds of folks, the common folks there, they taught me to stand up and insist upon things. That's how we learned to get things done for Sandy.

"I got into all kinds of shady dealin's while I was livin' in Maryland then. I don't mean at NIH but round Washington. I was buying things I knew were stolen. I went into this apartment once and the end of the kitchen cabinet could be moved out and there was a tunnel into the next apartment. It was filled with appliances and all sorts of things. You wouldn't believe it. They said, 'Mr. Hall, what would you like to buy?'

"I used to buy diamond rings of what you might call questionable ownership. I would take them to the pawnshop and leave them there a few days to see if they were hot. Then I'd get them out and sell 'em. Oh, I was a thoroughly unsavory character.

"In 1970, when Sandy was in her last months, I was drinkin' real bad. I drank mostly in the afternoon and at night so I could sleep and not have these nightmares. These dreams were terrible. They went on every night. In every one of 'em somethin' awful was happenin' to my family. They were fallin' down a hole or being crushed or burnin' up or somethin' terrible.

"I was in awful shape and I didn't know what to do. Well, son, Sandy, who was only twelve then, asked her pastor to come and see me. I didn't want to see no reverend, but she asked me to, so I said I would. He come to the apartment one night. I was desperate but I didn't know how to talk to him, what to say. I was scared of him.

"I finally worked up enough courage to tell him I didn't think I could go on. He just said, 'God can help you with that.' Then he turned away and went to talkin' to Sara and the young'uns. I was fit to be tied. Here I'd worked up the courage to confess that I had a problem and he just said, 'God can help with that.'

"But when he got through with Sara and the children, he came back to me. He had me get down on my knees and he opened his Bible and showed me where it told how God could help me. And I was saved that night. I really was. Sara was saved and so were the children. We became born-again Christians. We were saved in July. Sandy died in October. In November we went on the road singing gospel. I had never sung before, 'ceptin' in the shower. Now, Sara grew up in a family where they was always pickin' somethin'. Bluegrass mostly. So she could sing.

"That was in 1970. We've been singin' ever since."

Jerry begins to laugh. "I think the folks at NIH was happier to see me saved than I was. I was a thoroughly unpleasant fellow before I found our Lord."

A silence settles over us. Sara still sleeps. I turn and watch the white lines in the road disappear under the front of the bus. After a few minutes, Jerry speaks again.

"If I hadn't had God, I don't know what I'd 've done when Sue got sick. Lord, she was tough. I don't know how many times them doctors come to me and said, 'Mr. Hall, she's not respondin'. She doesn't have much longer left.' And then a couple of weeks later she'd be up and walk out of there. Of the two years she was sick she probably sang with us maybe fifteen months. We sang in Rockville, Maryland, on a Friday night and the next morning was when she went back into the hospital and her mind went. She only lived two weeks more.

"It was a psychosis she had that would come and go. The docs said they thought it was a hemorrhage in the front part of the brain, but I don't know. She would have periods when she was fine. Then she'd feel it comin' on. She'd say, 'Oh, it's comin' again.' Sue was determined to be able to describe what it felt like at the end. She wanted to tell us what she saw and felt as she died. She studied real hard on bein' able to do that.

"But she never got to. The night before she died we could talk to her up until about seven o'clock. She died at five-thirty the next morning. That was what God wanted."

I look at my watch. It is almost 2 A.M. We will be in Charleston soon. They plan to park near the airport, sleep until about six o'clock, and then drop me off to catch an early flight home.

"I wish you could've heard Sue sing," Jerry says wistfully. "Oh, that girl was good. A natural-born performer. You should've seen us together. If a crowd wasn't respondin' to me, she would take it and they would respond to her. We didn't even have to say anything. She'd just look at me and I'd look at her. She could get people to laughin' and then a minute later she'd have them cryin'. Then she could get them to laughing again. She was so good at testifyin' about God and his love for us and his forgiveness of our sins."

He sighs and leans back. "Lord, how I miss that girl. I miss her so much. We were so good together."

Later, I enter one of the bus's tiny bedrooms and lie down on one of the bunks and cover myself with a blanket. The bunks are six feet long, two inches too short for me to stretch out. The bus rocks gently and the whine of the wheels and vibration of the engine should put me to sleep. But sleep won't come.

I have never met a family like this. The intensity of their devotion to God and Jesus Christ is thrilling to me, even though I'm by no means a religious person. Their openness and the ease with which they draw me into their family is warming. Was there a time in America's history when more people oriented themselves outwardly like the Halls, rather than inwardly as so many of us now do? Their ability to bring a vividness to God and Christianity and share it with those who come to hear them sing is unquestionable. They have learned how to continue with life, to smile and accept what has been dealt them as well as cope with the uncertainty of the future. It is inspiring, whether or not you believe in God or Jesus or Christianity.

Soon the bus slows and I feel us turn off the interstate. We bump along a rough road and then all is still, except for the bus's engine, which is left idling to supply electrical power as we sleep. My thoughts take me back to the auditorium at the Bennettsville High School.

As The Singing Halls move into their final number, Jerry steps forward, his arm raised toward the heavens, his voice filled with emotion. The instruments play softly behind him.

"When we finish with our lives on this old earth, Sara and I have other children waiting for us. They're where the streets are paved with gold and no foot has ever trod before," he says, his voice filling again with tremulous emotion. "There is little Sammy. He's three and a half. There's Scott, who is eight. There's Sandy, who is thirteen years and four days old. And there's Susan. She's twenty-two."

His voice is brimming with real tears again. "God is keeping them for us. What better baby-sitter could you want than our Lord? Oh, how Sara and I look forward to the day when we can join them. When we will be reunited. It's knowing that they're there, waitin' for us, that keeps us keepin' on."

Steven begins to sing, his voice rising from deep within and filling the auditorium with its intensity.

Where the wilted flowers bloom again
in the garden of God
we will stroll together hand in hand
where feet have never trod.

It's the only thought that cheers me
and helps me make it through.
We'll shake hands in that land
where all things are new.

The Halls fill each chorus with more intensity than the last as they sing this song of promise of reunion with their dead children. Again, people in the audience come to their feet. Again, I see hands brush away tears. People begin to raise both arms above their heads, their testament to the glory of God and their belief in a better life after this one.

ANN AND RICK

No one knows who first thought of transfusing blood from one person to another. It is easy to imagine a primitive surgeon stumbling upon the idea on some ancient battlefield while watching a wounded soldier bleed to death.

In 1492 Pope Innocent VIII, in a coma and dying, was given blood taken from three young men. The young men died and so did the Pope. Historians suspect he drank the blood.

It wasn't until the twentieth century that Karl Landsteiner, an Austrian immunologist, discovered that human blood existed in different groups, or types. Landsteiner's discovery opened the door to blood transfusions—just in time for World War II. Doctors learned to match blood to be transfused with the blood type of the person receiving the transfusion.

Driving through Connecticut on a Sunday in November, I found myself thinking about the three young men who donated their blood to Pope Innocent. Did they volunteer in a moment of religious fervor? Had they been forcibly bled?

My destination was New Canaan, Connecticut, where a handsome couple in their thirties whom I will call Ann and Rick awaited me. They own an expansive, stylishly modern split-level home set back in trees on a large lot beside a golf course. The living room and dining room open onto a deck that overlooks a forest glade bisected by a small stream. We spent the afternoon in their living room nibbling on cheese, crackers, and nuts. As the sun sank beyond the hill behind their house, its light filtered through the bare tree limbs and cast an orange glow on the room's white walls.

Rick is tall and broad-shouldered and exudes quiet self-confidence. I had little difficulty picturing him as an executive of a New York City pollution control company. He studied chemical engineering at Cornell, where he was regarded as a serious, hardworking student whose ambition would ultimately carry him to success. He was born in New York City, but grew up across the Hudson in New Jersey.

Ann was raised in New York City, which her voice still betrays. She is the daughter of a policeman. Ann's mother died when Ann was two years old. She spent most of her childhood and adolescence shifted around New York's boroughs, living with relatives. She went to high school in Brooklyn, married in her early twenties, and had a son, Gregory. Ann was divorced and raising Greg alone and working as a secretary at the same pollution control company where Rick came to work.

After their marriage, Ann and Rick very much wanted to have a child. What they didn't know was that their differing blood types, when combined genetically, would make having a baby the most difficult, trying experience of their lives. It would be an experience that ironically might have been prevented if Ann had given birth to Greg just one or two years later. Ann is a victim of Rh incompatibility. Her babies can die in the womb of a condition called erythroblastosis fetalis.

This disease occurs once in about every 200 pregnancies. It's not a genetic defect in which unwanted genes are passed to children. Instead, it is a condition caused by unwanted antibodies in the mother's blood that react to genetically determined factors in the baby's blood. It affects second and subsequent pregnancies. Women with the condition can usually have one normal child before the trouble begins. Women who face this problem now are given an injection within seventy-two hours of the birth of each baby or after an abortion or miscarriage. If they receive the injection, the problem won't bother them. However, there are women like Ann who had their first child before the preventive substance was developed and are still of childbearing age. If they wish more children, as Ann did after marrying Rick, they are destined to have difficulties.

In Rh disease, antibodies in the mother's blood cross the placenta and attack the red blood cells of the growing fetus, destroying the baby's blood as quickly as it is manufactured. The result is usually a spontaneous abortion, a stillborn infant, or a seriously ill child who often dies within hours or days of birth. Children who survive to birth are given transfusions to exchange all their blood. Many go on to lead normal lives.

Most women with the condition must have the progress of their pregnancy monitored closely through blood tests that chart the buildup of the antibodies that could attack the baby's blood cells. If this buildup reaches dangerous levels, doctors may attempt to transfuse blood directly into the fetus to replace the blood being destroyed. A needle is inserted through the mother's abdomen and blood is injected into the fetus' abdominal cavity, where it is absorbed into the bloodstream. The procedure is risky and uncomfortable. But it often works. That turned out to be Ann and Rick's only hope. To receive this unusual treatment, they sought help from Yale University's high-risk obstetrical unit.

I wanted to write about a woman with Rh disease for several reasons. Solving the riddle of the disease is a fine example of research teamwork between geneticists and immunologists. Transfusing new blood into a growing fetus exemplifies the high technology now being used in obstetrics to deal with genetic problems. And the care that one of these babies must receive after birth in a newborn intensive-care unit challenges all the technology and expertise that has been brought to bear on the high-risk infant.

While attending an international gathering of geneticists in Montreal, I listened to the head of Yale University's high-risk newborn unit, Dr. Joseph Warshaw, describe progress in his specialty, neonatology. He outlined how younger and younger premature infants are being saved and predicted that there will be even more progress as newer technology and treatment is developed.

That night I attended a cocktail party for the conference participants held on two levels of the lobby of a large, modern concert hall. It was a loud, smoke-filled affair with doctors jostling one another at the bars and tables loaded with hors d'oeuvres. Warshaw is tall, and I spied him across the room standing with his wife. I shoved through the crowd and introduced myself. For some reason, he asked me where I lived.

"Copake!" he cried when I named the small village in upstate New York. "Is Camp Pontiac still there?"

"It's still there," I replied, surprised to discover yet another person who as a child had attended one or another of the summer camps in the area.

"I spent the most miserable summer of my life at Camp Pontiac," he confessed. "But it wasn't the camp's fault," he hastened to add. His wife was regarding him with an amused smile.

I quickly outlined my idea, shouting above the noise of the crowd. He was immediately enthusiastic and invited me to come see him at Yale when I was ready to begin. "I have just the case for you," he called, as the inevitable currents of the party pulled us apart.

Warshaw smokes a pipe and talks slowly, often flashing an ingenuous little boy's smile. He wears corduroy jeans and open-neck shirts and detests the traditional doctor's white lab coat. When I visited him in his office several weeks later, he was sprawled on a couch, his feet on a chair. His jeans were noticeably worn. He later confided, almost as a point of pride, that his secretary had once volunteered to buy him new pants when he turned up one day in a particularly tattered pair. He telephoned his colleague John Hobbins in the high-risk obstetrical unit down the hall and asked him to join us. I explained again what type of case I was interested in following.

"What about Ann and Rick?" Warshaw asked Hobbins.

"Perfect," the obstetrician replied without hesitation.

A few weeks later I was sitting in Ann and Rick's living room.

Ann has a snowy complexion and a round face framed by brown hair that gives her a delicate beauty, like that of a finely crafted china doll. She is energetic and speaks rapidly and enthusiastically, using her hands to tell a story. She radiates energy, and it makes her appealing. I could understand how Rick, who is deliberate and quiet, would be attracted to her.

"After I had Greg I had several miscarriages and apparently I was building up antibodies which would affect subsequent pregnancies, but I didn't know it," she said, sitting on the floor and leaning against the couch and Rick's leg. "Then in 1975 I became pregnant and things went well until the seventh month. I was going to a local obstetrician and he did some routine tests and discovered that there was an Rh problem. So he sent me to Yale for an amniocentesis and a scanning with ultrasound. That's how we met John Hobbins. The tests indicated that the baby might be affected, but not too severely. So my local obstetrician delivered the baby.

"I had the baby by caesarian section. It was only mildly affected with Rh and they would have transfused it and everything would probably have been okay, but the baby had a herniated diaphragm. Its intestines were coming through a hole in the diaphragm. They had delivered the baby early because my antibodies were building up. And because of this there was also the possibility of a problem with immature lungs. But with the hole in the diaphragm they had to rush the baby up to Yale and Rick jumped in the car and drove up behind the ambulance. That's when he first saw the newborn special-care unit at Yale. The baby died after several hours. Poor Rick had to wait through all that.

"The problem of Rh disease gets worse with each pregnancy. But that baby turned out to be only mildly affected with Rh and John Hobbins said he thought that if we tried again the Rh might not be a problem. We also talked about adoption, but that is so long-drawn and painful and difficult. We also considered artificial insemination. But in the end we decided to try again. I would be cared for from the first by the doctors at Yale. I had become a high-risk pregnancy."

Prior to 1940, infants were often struck by a group of diseases that had no explanation. The most striking symptom in both stillborn infants and aborted fetuses was a swollen appearance caused by excess fluids. The skin might be yellow, a sign of jaundice. Blood tests revealed large quantities of immature red blood cells. The liver, spleen, and other organs were often swollen.

In the 1930s, as medical researchers continued to probe the secrets of the body's immune system, suspicion grew that these infant diseases were immune-system-related. The work of Karl Landsteiner that led to discovery of the different blood groups formed the underpinnings of the research that solved this medical riddle.

Landsteiner found that human blood was divided into four basic groups, or types. They became known as type A, B, O, and AB. If blood belonging to Type A was mixed with blood of another type, B, for example, the red blood cells stuck together in clumps and became useless. If A was mixed with O or AB mixed with B, clumping occurred. But if like was mixed with like—A with A or O with O—nothing happened. Blood transfusions finally became possible.

Landsteiner and other medical researchers solved the transfusion problem because of basic discoveries about how the immune system works. The immune system is an exquisite defense system that readily distinguishes friend from foe, "self" from "nonself." When an outside invader, such as a bacterium, enters the body, foreign antigens are exposed. Immune system cells immediately detect these antigens and sound an alarm. Other immune system cells begin manufacturing substances called antibodies, which are like soldiers that fight the battle against the invader. The antibodies seek out the antigens and the invading organisms and bring about their destruction. Most amazing of all, the immune system has an incredible memory. Years after an invader appears and is driven off by the immune system, the immune cells remember. If that invader appears again, an immune system attack is marshaled again, but much faster and more strongly.

In 1940, continuing his experiments, Landsteiner and a colleague injected blood from a Rhesus monkey into rabbits. Antigens present on the monkey blood cells stimulated the rabbits' immune system to manufacture antibodies against the monkey blood cells. The doctors discovered that these antibodies would also react against human blood cells. Obviously there were antigens on the monkey blood cells that were identical to antigens on human blood cells. This unknown antigen was named Rh. Did it have something to do with the mysterious diseases of the newborn? The suspicion was that it did.

Some months earlier, other scientists had discovered that these newborn diseases were related to unknown antigens present in the blood of a fetus affected with the diseases. The antigens appeared in second and subsequent pregnancies of certain women. These antigens were named Antigen X. It wasn't long before more research revealed that Landsteiner's Rh antigens were identical to Antigen X. An answer was close at hand.

About 85 percent of all people have the Rh antigens in their blood and are Rh+. The other 15 percent don't have the antigens and are Rh−. Each person's blood also has antibodies that wait to attack other types of blood. For example, if your blood type is A, your blood carries antibodies that will react against blood cells from Types B and AB blood. If you receive a transfusion of Type B blood, the antibodies against Type B blood will go into battle, causing clumping of your own Type A blood cells as well as the Type B cells you received in the transfusion. That causes a severe reaction and often death. Thus there are always antibodies to most other blood types in your blood, waiting to destroy blood cells from another type.

But such antibodies aren't present where the Rh antigens are concerned. If you are Rh− and you are transfused with Rh+ blood, nothing will happen—the first time. But after the first transfusion your immune system has been alerted. The next time you receive Rh+ blood, antibodies to the Rh+ antigens will be waiting. There will be a reaction, possibly severe.

With that knowledge, doctors finally unraveled the mysterious newborn diseases. For the problem to occur, the mother must be Rh− and the father Rh+. Because of the genetic factors at work, the fetus growing in the mother's uterus would usually be Rh+. Red blood cells are too large to pass across the placenta from the mother's circulation to the fetus'. Thus, although the mother's Rh− antigens and the fetus' Rh+ antigens would produce an immune reaction if the two bloods were mixed, that doesn't happen during the first pregnancy.

But during delivery of the first child (during the birth of Greg, in Ann's case) some of the infant's blood enters the mother's body, probably as the placenta is removed. If the infant's blood type is different from the mother's, say Type B in a mother with Type A, then the mother's Type A antibodies immediately attack the infant's wandering Type B blood cells and destroy them. There will be no Rh problem with future pregnancies for that woman. But if the infant's Rh+ blood is of the same type—say, Type A in the mother and Type A in the infant—then the infant's blood cells won't be destroyed. They will remain. But the infant's Rh+ antigens will cause a reaction by the mother's immune system. The infant's Rh+ antigens stimulate the production of anti-Rh+ antibodies by the mother's immune system. Her immune system is now sensitized to Rh+.

During the next pregnancy, the anti-Rh+ antibodies in her blood are small enough that they *can* cross the placenta to the fetus' bloodstream. As the fetus grows, these antibodies cross over in increasing numbers, searching out and attacking the fetus' Rh+ blood cells. To compensate, the fetus' blood-forming machinery begins to work overtime, enlarging itself, but to no avail. The anti-Rh+ antibodies in the mother's blood are relentless. If they don't eventually kill the fetus, causing a miscarriage or a stillbirth, then the baby will be severely ill at birth.

Moreover, this situation worsens with each pregnancy. More and more of the anti-Rh+ antibodies are in the mother's blood, waiting. It is as if an army platoon has opened fire on its fellow soldiers and refuses to believe a mistake has been made.

Ann had Type A blood and was Rh−. Her first husband was Type A and Rh+. Greg was an Rh+ baby and a few drops of his blood entered Ann's bloodstream when he was born, causing her body to form anti-Rh+ antibodies. Rick, in turn, was Rh+. All of Ann's subsequent pregnancies carried an Rh+ fetus, and her anti-Rh+ antibodies crossed the placenta and attacked the fetus' Rh+ blood. It had nothing to do with the hole in the diaphragm of the baby born in 1975. But it did account for the mild Rh disease the doctors detected in that baby before it died.

If doctors know that an Rh− mother has given birth to an Rh+ baby of the same blood type, all they need do is immunize the mother within about seventy-two hours of the first birth, thus preventing antibodies from being formed. But development of a material that could be injected to accomplish this didn't come until the early 1960s. It wasn't in widespread use until one or two years after Greg was born.

During the final weeks of her 1975 pregnancy, which produced the baby with the herniated diaphragm, Ann went to Yale every two weeks for an ultrasound scan and amniocentesis to measure the extent to which the Rh problem was affecting her growing baby. Rick would leave work to accompany her.

One day as Ann lay on the treatment table while John Hobbins was preparing to insert the amniocentesis needle into her abdomen, a white-coated doctor entered the room. Neither Rick nor Ann noticed his face. He stood with his back to them, conferring in low tones with Hobbins.

Rick studied the doctor's back. Slowly he began to realize that something was familiar. It was his build, the way he stood. Suddenly it clicked. "Is that Dick Berkowitz?" Rick blurted.

The doctor whirled. It *was* Dick Berkowitz. There followed one of those unexpected college classmate reunions filled with handshakes and exclamations of surprise, subdued somewhat by the surroundings and the medical procedure under way.

Dick Berkowitz and Rick had both belonged to Tau Delta Phi social fraternity as undergraduates at Cornell. They lived in the frat house and were friends, although not extremely close. Seeing Berkowitz at Yale was particularly startling to Rick. His fraternity friend had been a cutup who preferred parties to study, brilliant in those courses that interested him and squeaking through the rest. It seemed amazing that Dick Berkowitz had become a doctor, let alone a member of the obstetrics and gynecology faculty at Yale. In fact, Rick learned, Berkowitz was such a brilliant obstetrician that he had been invited to join the faculty and team with Hobbins in the high-risk obstetrics program.

All during the latter part of her 1975 pregnancy, Hobbins cared for Ann. A year after that baby died, Ann became pregnant again. This time, Hobbins and Berkowitz were to care for her jointly, as they do with all their patients. But when Ann's troubles began late in the summer of 1976, Hobbins was on vacation. The full responsibility for her care fell to Berkowitz, including the risky intrauterine transfusions. The fact that he had renewed his friendship with an old fraternity brother and now had exclusive responsibility for the care of his friend's wife ultimately presented the doctor with one of the more difficult ethical crises of his career and caused him considerable personal distress.

"Ann is a person who was put on this earth to be a mother. That's what she's wanted from life more than anything else," Berkowitz told me admiringly when we first talked. He and Hobbins have adjoining offices separated by an anteroom where their secretary works. The offices are directly across the hall from the treatment room where they perform the amniocenteses. It was here that we spoke. Berkowitz' office is small: a couch along one wall, the opposite wall lined with books, a metal desk, and a swivel chair that rocks back. He still wore surgical greens when we talked, his arms clasped behind his head as he leaned back in the swivel chair. He has a mustache and curly hair. I often saw an impish gleam in his eyes.

Berkowitz said that his father was a doctor and that as a college student he started out by rejecting medicine. "I went to the same elementary school, the same junior high school and senior high school as my father," Berkowitz said, crossing his legs and leaning back until I wondered if the swivel chair might topple over. "I was absolutely certain that if I went to the same college and into the same profession my life would be a carbon copy of my father's. So I decided not to become a doctor. My father went to Michigan. I went to Cornell to become a chemical engineer. But I called home after the second term began and said I wasn't interested in being an engineer. I wanted to go into liberal arts. My father said that was fine, but that I shouldn't just be 'in' liberal arts, I should have a goal. He said if I changed my mind about what that was, that would be okay, but that I should be heading toward something. That forced me to do a lot of thinking about the kinds of things I enjoyed doing and how I wanted to spend my life. Eventually I decided medicine fit the bill. So I went to medical school at New York University. My family has its roots in Lithuania where it held the hereditary post of doctor at a prison. As far as we can trace back, I'm a sixth-generation physician."

Berkowitz spent two years as a Peace Corps physician in Africa. After finishing his postgraduate training as an obstetrician and gynecologist, he came to Yale for a summer to learn some anesthesiology before returning to Africa. He wanted to conduct gynecological surgery in Africa with just the aid of a nurse, performing surgery and administering anesthesia himself.

But he impressed Yale so much they offered him a position on the faculty. He was steadfast about his desire to return to Africa. Yale agreed to wait for him. After spending fifteen months in Tanzania, Ethiopia, and Kenya, including six months at a hospital in Kenya where he was in charge of obstetrical and general surgery, he returned to Yale.

Rick had told me about his impressions of Berkowitz as an undergraduate. When I asked Berkowitz about his undergraduate days, a sheepish look crossed his face.

"Rick was and is a quiet guy," he said. "He is a solid citizen. Very smart. A really good student, but also well rounded. A good athlete. Just a nice guy.

"When I was in college, I fooled around a lot. I enjoyed myself. I worked hard in the courses I liked and not so hard in some of the others. But I got good enough grades to get into medical school. Rick and I lived in the same fraternity house for a year when I was a junior and Rick was a sophomore. I think that when I turned out to be Ann's doctor it took some adjusting for Rick to realize that this was the same individual he remembered from college. He's said several times since that it is hard for him to believe I was taking care of his wife. But I've changed. At least, I hope I've changed."

An obstetrician gets the first indication that a pregnant Rh− woman's fetus may have problems when blood tests on the mother reveal increasing levels of anti-Rh+ antibody. Something is stimulating the woman's immune system to manufacture these antibodies. The assumption is that the growing Rh+ fetus is responsible. But blood tests alone can't reveal the extent of the trouble. Doctors must sample the environment in which the baby is growing.

Amniocentesis was developed in the 1960s to deal with this problem. The destruction of the fetus' red blood cells by the antibodies coming from the mother's immune system causes a buildup of a substance called bilirubin in the fluid surrounding the fetus. Charting this bilirubin buildup, as well as examining the fetus with ultrasound to look for swelling, will accurately show the progress of the Rh disease.

When Ann had been pregnant for twenty-two weeks she went to Yale for the first amniocentesis of her new pregnancy. As expected, the levels of anti-Rh+ antibodies in her blood had been rising. It was almost certain that Hobbins and Berkowitz would find elevated bilirubin levels. The ultrasound might reveal other difficulties.

"There is some bilirubin present in the amniotic fluid of a normal pregnancy," Berkowitz said, leafing through Ann's medical records folder. "The amount is generally a function of the gestational age of the fetus. On a piece of graph paper you plot the amount of bilirubin against the number of weeks of gestation. The curve that results will fall into one of three zones—Zone 1, 2, or 3.

"Zone 1 means the baby is totally unaffected with the Rh problem or only mildly affected. Zone 2 is the transition zone. When someone moves into Zone 3, that means trouble. You have three options. First, you can do nothing and the fetus will die. Second, you can deliver the baby immediately by caesarian section. If the baby is only, say, twenty-eight or thirty weeks old, it may die of problems associated with prematurity. The third option is intrauterine transfusions."

I had read up on such transfusions. I knew they were risky for the baby and uncomfortable for the mother. A large-diameter needle is inserted into the mother's abdomen and on into the fetus' abdomen. Blood that has been carefully checked for its blood type and antigens that might cause the mother's immune system to mount an attack is injected into the fetus' peritoneal or abdominal cavity. From there it is assimilated by the fetus' blood-starved tissues.

Sometimes, when such an intrauterine-transfused baby is born, its body contains 90 percent or more transfused blood. The transfusions are a delaying tactic, a buying of time. As the pregnancy advances, the mother's anti-Rh+ antibodies grow stronger and continue to attack the fetus' blood-forming tissues. The transfusions keep the fetus growing until it is big enough to be delivered by caesarian section and has a chance of survival in the newborn special-care unit.

In the last weeks of pregnancy, Berkowitz and Hobbins confer frequently about the Rh cases, watching the rise of bilirubin and studying the ultrasound results and other tests. They wait until the last possible moment to take the baby from the womb. They are like the high-wire aerialist's assistant, watching the maestro teeter on the wire, but not wanting to grab him until there is no doubt he is about to fall.

As the afternoon sun dipped behind the hill at the back of Rick and Ann's home and the living room filled with shadows, Ann continued her monologue, still resting against the couch and Rick's leg. Occasionally she rose to pace the room or offer more cheese or a soft drink. But she always returned to her spot on the carpet. Realizing at one point that her words were tumbling out in what seemed like a cathartic avalanche and perhaps sensing Rick's affectionate and amused smile, she turned to him and asked, "Do you want to tell some of this? You lived through it, too. Am I talking too much?"

He smiled and shook his head. "You're doing very well," he answered.

Ann patted his leg and resumed. "They started the amniocentesis taps at twenty weeks. John Hobbins was doing the taps because he'd done them in the previous pregnancy. The first two taps, the bilirubin was in low Zone 2. The baby was mildly affected, but they weren't too worried. The baby wasn't in trouble or anything.

"I went up for the third tap one morning and John Hobbins was on vacation, so Dick Berkowitz did it. It was routine and I came back home. Then in the afternoon Dick called and he was very concerned. The lab analysis of the amniotic fluid showed that I had to have an intrauterine transfusion. He wanted me to come up that night so he could do the transfusion first thing the next morning.

"I remember this as one of the times I got very emotional. I started crying. I called Rick at work and he came home and we drove right up to Yale. I was crying on the way up and I said to Rick, 'Maybe these are extraordinary means that we are going through, beyond what we should be doing. Maybe I shouldn't go. I could refuse the transfusions.'

"We met Dick in his office. He suggested that perhaps we might want to go to Boston to Children's Hospital to have these transfusions. He was very frank about it. John Hobbins was on vacation and Dick hadn't done that many intrauterine transfusions, at least not alone. He said just because he was a friend that we wouldn't be hurting his feelings if we decided to go to Boston."

I had noticed Rick fidgeting as she said this and I wasn't surprised when he broke in. Had his undergraduate image of Berkowitz intruded?

"It wasn't hard to make a decision," Rick explained. "The fact that I knew him in college wasn't a factor. We knew that Dick is extremely capable. That comes across very quickly. We had gained a lot of confidence in him. He obviously had confidence in himself. Plus, we would have to travel to Boston. And there were likely to be several transfusions. So we told him to go ahead."

Later, I asked Berkowitz why he suggested that Rick and Ann go to Boston.

"I had done some intrauterine transfusions, but not too many," he said slowly. "I hadn't done any when John wasn't available. He might not have been in the room but he was always around in case there was a problem. I told Ann and Rick that the transfusions were something that could be done by me, or they could be done in Boston. But they had to be done soon. If they wanted, they could go to Boston. I had telephoned that afternoon and made all the arrangements.

"I was very concerned about my personal involvement. It was something that had to be dealt with. I felt they would be in an extremely awkward position about exercising options if something did go wrong. I didn't want them to feel I had sealed them into an optionless corner. I wanted them not only to have the best care possible, but to *feel* they were having the best care.

"They sat right on this couch and made their decision. They were obviously comfortable with staying at Yale . . . with me doing the transfusions. What I didn't tell them was that I secretly hoped they would go to Boston. I still felt very uneasy about doing the transfusions. I didn't sleep at all that night. I worried about it all night. I had never encountered feelings like this as a doctor before."

Ann spent the night in the hospital and the next morning was taken to a treatment room equipped with an X-ray unit and an ultrasound scanner. Berkowitz used the ultrasound picture to locate the baby's abdomen. As was usually the case with Ann, the fetus was in an incorrect position and she had to get up on all fours while Berkowitz tugged and pulled at her belly, trying to move the fetus. He had no luck, so he sent her out to walk the halls. When she returned, the baby had moved and the abdomen was exposed for insertion of the needle.

Berkowitz gave Ann a local anesthetic just below the navel and inserted the first needle, using the last ultrasound image to guide his thrust. When he thought the needle was in place, he injected a small amount of opaque dye, which shows up as a dark spot on an X ray. The technician took X rays and hurried into an adjacent darkroom to develop them while everyone waited, the needle protruding from Ann's stomach. Berkowitz sighed with relief when the technician returned with the films. The needle was in the abdominal cavity—perfectly placed.

The doctor then attached an apparatus to the needle that permitted drawing blood from a bag hanging overhead, injecting it through the needle, and monitoring the buildup of pressure in the baby's abdominal cavity as more and more blood was pumped in. Through it all, the single needle protruding from Ann's abdomen danced and jiggled as the baby moved. Little is known about whether fetuses feel pain, but the assumption is that the needle does cause a sensation, perhaps painful, which causes the fetus to thrash about.

Berkowitz injected 60 cubic centimeters of blood that day, a satisfactory amount. Ann came back a week later for a routine amniocentesis to monitor bilirubin levels, which were stable, reflecting the benefits of the transfusion. A week after that, she returned for the next transfusion.

When Berkowitz ran the ultrasound scanner's transducer over Ann's belly preparatory to giving the second transfusion, the picture that flashed on the television monitor was a chilling sight. Fluid had begun to build up in the baby's tissues and the swelling was obvious on the screen. It was a sure sign of approaching heart failure, a condition called hydrops. The fetus' tiny heart wasn't beating strongly enough and couldn't push enough blood through its blood vessels to keep all the body's chemical processes in balance. Excess fluid was the result.

The fluid buildup was a grave sign. Only 15 percent of all Rh fetuses with congestive heart failure survive. Berkowitz went ahead with the transfusion, but he had to insert four needles before he finally had one in the right position. He was able to inject only 49 cc of blood, a disappointing amount.

Ann returned to her hospital room, as she did after each transfusion, and Berkowitz called Joe Warshaw. He explained the problem. "She's at twenty-nine and a half weeks," he said. "What would happen if we took the baby now? What if I did a C-section now?"

There was a pause at the other end of the line. The baby would have a 10 percent chance of survival, came the answer.

Berkowitz called a pediatric cardiologist on the Yale faculty and asked his advice. The best treatment, the cardiologist said, might be to give Ann digitalis, a heart stimulant. The drug would cross the placenta to the fetus' blood and possibly stimulate its flagging heart. The drug wouldn't have an adverse effect on Ann. But it might make it possible for the baby to make use of the newly transfused blood. Berkowitz met with Ann and Rick to outline the problem.

"I told them that we could see the fluid and what that meant, that the statistics were pretty grim. I told them we could take the baby, but that the odds were poor. I outlined the digitalis treatment and told them that was what I wanted to do. All we could do was give Ann the digitalis and wait and see.

"There's a fine line here between giving the patient all the facts and letting them help make decisions but still managing the case yourself. This kind of decision can't be an emotional thing. You hope you have the very best information possible and then you couple that with your judgment and your experience. Medicine is often more art than science, but it's an art based on experience."

The gamble paid off. When Ann came in for her third transfusion, the ultrasound showed the fluid had gone. Presumably the digitalis had stimulated the fetus' heart sufficiently.

But it was a difficult transfusion, nevertheless. Berkowitz inserted one, two, three needles, waiting each time for the X ray to come back. Each time the dye revealed that the needle wasn't on target. Ann waited, lying on the table, her discomfort and her anxiety growing with each needle, watching the needles dance and jiggle where they protruded from her belly.

"Once the first needle misses the mark, it gets tough," Berkowitz said, grimacing at the memory. "Interpreting the X rays is hard. They are two-dimensional and they show the needle and the fetus, but they don't show the angles involved. Ultrasound shows us the angles, but then you have to turn it off to actually insert the needle. If the baby moves in the meantime, then it's all changed again."

The risk to the fetus during intrauterine transfusions runs as high as 10 percent. The big danger is that a needle will hit a vital organ, the heart for example, and cause serious damage, even death. There is always a danger of an infection in the womb that could cause a spontaneous abortion.

When Berkowitz inserted the fourth needle that day, the danger that always loomed suddenly struck. When the technician hurried back into the room, the films showed the dye spreading to the fetus' bowel. He had inadvertently shoved the needle too far and pierced the intestine. He backed the needle off slightly. Concentrating intensely, sweat pouring from his forehead, he injected more dye to see if pulling the needle back had placed it correctly. When the next X rays came back, he relaxed a little. Finally, the baby had paused in its movements and the needle was on target. He began injecting the blood.

After Ann returned to her room, Berkowitz met with her and Rick and told them about piercing the bowel. He said there was no way to tell what damage, if any, had been done. He asked them to come back in two days for an ultrasound scan. It revealed that everything was fine. There had been no short-term damage.

When Ann and Rick talk about this period, they are matter-of-fact. The memories of the anxiety and the intensity of emotion have inevitably been eroded by time. Berkowitz, however, retains a sensitive perspective.

"The kind of work we do here with high-risk obstetrical patients and that which Joe Warshaw does in his unit is very emotionally charged stuff," Berkowitz said. "We're dealing with difficult problems in people who care an awful lot because they are willing to go to great extremes to have a baby. There's a great amount invested in their case on everyone's part. But we don't always win. Sometimes we lose. That creates a lot of pressure. I think a lot of people in medicine wouldn't like that pressure, and none of us like it when we're in the middle of it. The pressure's tough, but the rewards are the greatest."

Two weeks after the fetus' bowel was pierced, Ann appeared for her fourth transfusion. Berkowitz knew that with each passing day the chances of something going wrong increased. Ann's immune system was determined to get rid of the fetus, this seeming cancer growing in her uterus. There was no chance the baby could remain in the womb for a full nine months. It was only a matter of time before heart failure would set in. The trick for Berkowitz—the melding of the science and art of medicine—was to wait until the last possible moment and then deliver the baby. The critical factor became the maturity of the fetus' lungs. If the baby was delivered and the lungs weren't ready, it would develop respiratory distress syndrome in the newborn special-care unit. That condition, combined with Rh disease, is often fatal.

Thus the amniotic fluid withdrawn every two weeks took on another importance: measurement of fatty substances that indicate lung maturity. The substances are known chemically as lipids and are named lecithin and sphingomyelin. A fetus' developing lungs produce both of these lipids. Before the thirty-fourth week of gestation there is more sphingomyelin in the amniotic fluid than lecithin. This means the lungs are still immature. At about thirty-four weeks lecithin production jumps markedly. When there is twice as much lecithin as sphingomyelin, or a ratio of 2 to 1, doctors are assured that the lungs are mature enough for delivery. There is little likelihood of the newborn child developing respiratory distress syndrome. A ratio lower than 2 to 1 often means trouble. When Ann received her fourth intrauterine transfusion at thirty-three weeks' gestation, the ratio was disturbingly low.

"It was 0.6," Berkowitz said, studying Ann's medical records. "That meant the lungs were extremely immature and it seemed certain the baby would have respiratory distress syndrome." He pawed among some papers on his desk and extracted a document. "For example, just this morning I delivered an Rh patient. She is thirty-three weeks and her ratio was 1.7. So I chose to deliver her rather than transfuse her again. That's what I would have preferred to do with Ann: deliver her at thirty-three weeks.

But I just couldn't do it with a ratio of 0.6. So we went through a fourth transfusion. The first needle was right on target and we had no problems.

"Then she came back a week later for another amniotic tap. The bilirubin was still high, which didn't surprise anyone. But the lecithin to sphingomyelin ratio had dropped! It should have risen but instead it had dropped to 0.4. The low L to S ratio was disappointing, but all other signs were stable. She didn't need to be transfused again for another week. So we decided to wait another week.

"A week later she came back and we tapped her again but didn't transfuse her. We wanted to wait for the lab results on the ratio. But it was still 0.4. At this point we had to either transfuse her again or deliver the baby. We were reluctant to transfuse again because of the inherent risk to the fetus every time you do that. So we decided to give her steroids to attempt to speed up lung maturation.

"Here's the art side of medicine again. At thirty-three weeks we felt the baby would do better with another transfusion than with delivery. But at thirty-five weeks we felt it would do better in the nursery, despite the low lecithin to sphingomyelin ratio and the risk of respiratory distress syndrome. But to cover our bets, we gave her steroids and got ready to deliver the baby by C-section.

"We don't use steroids in our practice very much because they are potent drugs with many side effects. But at that time there were several medical centers in this country giving steroids to stimulate fetal lung maturity. Our concern was that the steroids could affect brain development in the fetus. There had been some animal research suggesting that. But no one had come up with any evidence of problems yet. And it was such a tough situation with Ann. Deciding to use steroids was a big decision for us. But the risk from steroids seemed lower than the risk of another transfusion. We just had to get that baby out of there. We wanted to maximize every chance."

The decision to administer steroids was made on a Saturday in October. The next day Ann returned to Yale for an injection of steroids. She returned again Monday and Tuesday for more injections. Late Tuesday afternoon she checked into a room on the obstetrical floor of Yale–New Haven Hospital. Berkowitz and an assistant would deliver the baby the next morning.

Ann and Rick waited for Berkowitz in the hospital room. They had made plans to go to dinner together. He sauntered into the room early in the evening, announcing he had just come from the library.

"Looking up how to do a C-section?" Rick joked.

They went to a seafood restaurant.

"Dick ordered wine," Ann remembers. " 'Better watch it over there, fellow,' I told him. 'You better stay sober over there.' He ordered a big meal and made me order one, too. Something with octopus. 'Dick,' I said, 'should I be eating all this?' He told me to eat. He drove us back to the hospital. I felt like I was in a hotel. It was late and I had to ring a bell at the door so they would let me in. 'See you in the morning,' Dick said. He gave me a kiss. 'You'd better, mister,' I told him."

Ann remembers two anesthesiologists. "The head one was a woman and she was explaining things to the other one. I got a needle before I went in and then she told me they were going to give me a spinal. I almost cried at that. I imagined all this pain. But I didn't feel a thing. She was lovely throughout. She sat by my head and talked to me. I started telling her about the other baby, the one that died. That must have been on my mind. She just kept smiling and smiling at me. I was talking to her and watching Dick. He had a very serious face. It was the same serious face that I had seen during some of the rougher transfusions.

"Dick asked about my scar. What kind of scar I wanted. With the other baby I had a scar that was horizontal across my abdomen. Dick wanted to make one that would be vertical. He said I wouldn't be able to wear a bikini. I told him I didn't care one bit. Just get that baby out of there.

"Then I started to cry. I don't know why. Dick was looking at me crying. 'You shouldn't be crying now. This is a happy time,' he said. That just made me cry more. The thing that was on my mind was that the other baby never cried. He just made gurgling sounds. I never heard him cry. Then I noticed there were other people in the room, sort of standing around. They had on surgical greens. I kept wondering who they were. And then I realized it was the pediatricians from the perinatal unit. Joe Warshaw's people. They seemed to have a table in the corner and they were laying out their things. They were waiting to take the baby down to their unit.

"Then the baby came out. The first thing someone said was, 'Oh! It's a boy.' I heard Dick say, 'My, he's big.' I kept listening for him to cry. But there wasn't a sound. 'Isn't he going to cry?' I asked. 'Isn't he going to cry?' I don't know how many times I asked that. But no one said anything."

Berkowitz has performed many caesarian sections and the details of each one tend to merge. But Ann's he remembers, probably because the patient also was a friend.

"The C-section was routine," Berkowitz said. "Ann received spinal anesthesia. We draped her and washed the abdomen with a sterile solution. She was awake, so I could talk to her. She was comfortable, but she felt nothing. There was a heart monitor on her—that's standard—and it was beeping in the background. With a spinal the woman is literally paralyzed from the waist down. She doesn't do any pushing or anything to help deliver the baby.

"The biggest decision we had to make there in the operating room was about her incision. If there is no rush, you can do what is called a bikini incision so it won't show as much. But Ann already had had two C-sections and had a large horizontal scar. I didn't want to go through that scar because there might be a lot of chipping away at tough, fibrous tissue. I might end up with a compromised incision. I wanted the delivery to be effortless because I knew the baby was stressed. So I told her I would have to make a vertical incision. Her scar would look like an anchor. But she couldn't have cared less.

"We opened the abdomen in layers and entered the peritoneal cavity. I put in a retractor, which pulled everything back and exposed the lower segment of the uterus. Then I picked up the peritoneum, which is a tough, smooth, and colorless membrane sac that holds the abdominal organs, and snipped it open. I dissected it away and pushed it down, which pushed the bladder out of the way. Then I made an incision in the uterus and lifted the baby out.

"I was holding the baby by its legs and head. The resident

clamped the umbilical cord and I turned around with it and a pediatrician was standing there with a sterile sheet draped over his arms. I literally dropped the baby into the sheet. They use a sterile sheet so I remain sterile. I still had a lot of work to do.

"It was a big baby, which surprised me. You get used to seeing these little pipsqueaks when you deliver Rh babies. There was no evidence of hydrops—the fluid buildup. The belly wasn't swollen and the placenta didn't look water-laden. And there was no evidence of any immediate acute problems.

"Ann was terribly, terribly concerned that the baby was okay. She was crying and I could see the tension in her was enormous. She was saying something about she couldn't hear him crying. They were working on him in the corner and then he began to cry. Everybody let out a sigh. Sort of a little cheer. Then the pediatricians were getting ready to zip him out of the room and someone asked if they weren't going to show him to his mother. So they held him up to Ann's face for a few seconds and then they were gone.

"There was a tremendous release of tension then. We gave Ann a big jolt of narcotic to make her all snowy and went back to work. She was out of it for the rest of the day and had an uneventful recovery."

Rick had stayed with Ann in her room on the obstetrical floor until the orderlies took her away for the delivery. Then he wandered to a waiting room near the newborn special-care unit. He didn't go in. He had been in the unit with the other baby, the one that was rushed to Yale, only to die. The rooms full of the high-technology perinatal equipment were a painful memory. During one of their innumerable trips to Yale for amniotic taps or transfusions, he and Ann had received a tour of the perinatal unit from Joe Warshaw. As they left the intensive-care room, a large room filled with incubators and noisy with the beeping of heart and other monitoring systems, Ann paused at the door.

"Is this where they brought our baby?" she asked.

Rick nodded and pointed to a corner of the room. "He was over there."

Ann stared in the direction Rick pointed for a few seconds and fought back a sudden urge to cry. Then they moved on in silence, following Warshaw's loping figure.

"After they delivered the baby, someone from the perinatal unit came and found me in the waiting room," Rick said. The shadows were growing deeper in their living room and Ann rose to snap on lights. Greg had come home from an afternoon outing and slipped into the room to listen. "They took me into the same room where the first baby had been. They were standing around a heated table. They had just brought the baby from the operating room. They were already getting ready to begin transfusing him with blood, to exchange all the blood he was born with for new blood free of Ann's antibodies.

"You remember funny things. One of the interns or the resident, a woman, had on jeans and combat boots or hiking boots. They seemed so casual and yet so intense. They were hooking up lines to the stub of his umbilical cord. Someone said, 'Come on over and welcome your son into the world.' I was amazed. They were telling me to touch him. Just to reach out and touch him. I put my hand in and touched a finger to one of his hands. He kind of moved his fingers. It was hardly possible to believe he was here. I felt like I already knew him. We had been through so much together."

BABY BOY SMITH

The condition of Baby Boy Smith worsened during the night.

The nurses noticed it first. He didn't respond in the same way he had a few hours before. They had cared for enough premature and gravely ill newborns to recognize the signs of neurological deterioration.

When he arrived in the morning, the resident looked in on the Smith baby first. He conducted a quick neurological exam—checking the baby's ability to grasp a finger, pinching a foot. The attending—the senior physician in charge of the unit—examined the baby when he arrived and agreed with the resident's assessment. They ordered a computerized axial tomography scan. It revealed that the nurses were correct. During the night, a massive hemorrhage had destroyed most of the infant's brain. Baby Boy Smith, if he survived, would be hopelessly retarded.

My intention had been to write about Ann and Rick's ordeal in Yale's neonatal special-care unit with their new son, Brian. I wanted to use their story as a means of describing how neonatology has developed in the last twenty years, becoming a court of last resort for genetic mistakes and disasters. Neonatologists, using the best of twentieth-century medical technology, often perform medical miracles, rescuing infants once doomed to die. They certainly had done that with Brian, rescuing him from the brink of death more than once.

So I drove to Yale one cold but clear winter morning and sat in Joe Warshaw's office, warmed by the sun streaming through his windows and eyeing the high stack of file folders on Warshaw's coffee table. They contained Ann's and Brian's medical records. My plan was to go over the medical records with War-shaw and then spend the rest of the day in the neonatal special-care unit itself, watching the daily routine. I wanted to get the feel of the unit and the people who work there. Warshaw was dressed in his usual corduroy jeans but wore a tie with his sports shirt. He closed the door so the noise of his secretary's typing wouldn't interfere with my tape recorder and then sprawled on the couch, his feet on the coffee table.

Several years before, I had participated in a seminar on medical ethics in which a film about a baby left to die in a hospital nursery was shown. The infant had been born with a serious birth defect. A long series of operations could ensure the baby's survival, but it would spend its life grossly retarded both physically and mentally. The doctors and the child's parents decided against the surgery. The infant was made comfortable and moved to a far corner of the nursery and left to die. The film was very moving and triggered an emotional debate afterward, as its makers doubtless intended. This was before the idea of ceasing heroic measures to save a life had come so much to the public consciousness because of the publicity surrounding the Karen Anne Quinlan case. Now people seem much less shocked by the idea of turning off the machines. I knew that in every special-care unit such decisions must be made several times a month, perhaps even weekly. I wanted Warshaw to talk about this.

"We find that most pregnant women, and the fathers, too, have fantasies and fears during pregnancy about the possibility of something going wrong," he said, filling a pipe with tobacco and lighting it. The smoke rose up into a shaft of sunlight from the window. "I know my wife had those fears. When the lightning strikes out of the blue they aren't so surprised. We here have an attitude that we shouldn't use technology to support a horrible quality of life. And when we talk to families about this we find that in most cases they already have thought about it. They are prepared to deal with it in a way that physicians often don't anticipate. They don't need to be protected as much as physicians and the general population seem to think.

"Having a child with a major genetic problem can be an awful family liability. I think the best example is a child with a major spina bifida, which causes severe mental and physical handicaps.

Some families end up coping with such a child very well. But others, if you were to give them a choice ten years later, would frankly wish that their child had not lived. It ends up being such a financial and emotional burden. Rick and Ann are one of our genetic success stories. They had a genetic problem but one which didn't leave any scars, unless there are emotional scars later.

"When you work in this field you have to learn to look at things in a fatalistic manner; at least, I have to on my own personal level sometimes. You have to be fatalistic to care for these cases. An extremely premature baby, say an 800-gram [28-ounce] baby, can engender a fantastic family attachment, a tremendous nursing attachment. We will do all that we can for such a baby until we have signals that we are doing too much, that nothing more can be done. If we don't win, and there are many times that we don't, then it's not the same quality of loss of life that is present if a three-year-old dies or a twenty-year-old. That's my emotional feeling and I'm sure not everyone agrees with that.

"We have a mother in the unit now who is very expressive, very vocal. She had a 1,000-gram [32-ounce] baby by C-section at twenty-seven weeks' gestation. Five years ago, if she went into labor at twenty-seven weeks, the obstetricians wouldn't even have tried to save it. Now they do. That baby has sailed right through. The couple wanted this baby very badly because they had lost a previous baby. It was a third marriage for him, first for her. For them it was a very high-priority pregnancy. She has been literally living here for six weeks, even providing breast milk for her baby. The attachment they now have may never be greater in the life of this child. From the physician's perspective that's wonderful. We would have felt terrible if that baby had died after a few days. But within the big picture of things, if the baby had died, its death to me would not have been qualitatively the same as the death of a three- or four-year-old."

At this point, Warshaw's telephone began to ring in his secretary's office. I could see the lighted button on the instrument on his desk flashing. It rang several times and he was about to straighten up and reach out for the receiver when his secretary answered it.

"Do you play God sometimes when you're making these decisions?" I asked.

"Do we play God?" he echoed, wrinkling his brow.

The intercom on the telephone buzzed. He turned and frowned at it. I saw him decide mentally to ignore it. He was hoping his secretary would take a message. "I don't think we play God," he answered slowly.

Suddenly there was a knock at the door. "Excuse me," she called through the door. "They're on the phone with a medical emergency."

I have never seen someone answer a telephone so quickly. With one lightning motion he straightened up, grabbed the receiver, and barked, "What's up?" In an instant he had become as taut as a cat gathering its haunches to spring, every muscle tensed. I expected him to literally slam the telephone down and run from the room. I was wondering if I dared follow if he didn't invite me.

But after listening intently for a few seconds he began to relax. I had switched off the tape recorder in deference to the privacy of the call. But he made no attempt to conceal his end of the conversation.

"Are you comfortable with that?" he asked the caller. "What about the nurses?" he said later. "Have you talked to the family?" He listened some more. "We'll talk about it later," he said, hanging up.

"Now, where were we?" he asked, turning back to our conversation. He smiled apologetically. "A medical emergency, but not something requiring immediate heroic action," he said, apparently feeling a need to explain the call.

Naturally I was bursting with impatience to ask him what was happening. "I asked if you ever felt you were playing God," I said. I hoped that as we talked he might begin to reveal what was happening at that moment in his unit.

"No, I don't think we play God. We try to have families involved in the decision making as much as we can. Of course, families are going to respond to the information you give them and we control the information. So that's always a hazard. But I think we make the best attempt we can to pick up on what the family really wants to happen. And then we respect that. What we can tell them is what we would do ourselves, what our medical judgment is for a given situation. An example would be this medical emergency going on right now. . . ."

I felt a stab of elation.

"We have a baby who has been on a ventilator, on artificial respiration, for two days. The baby has chest tubes, which are hoses we insert through the ribs into the chest cavity to create a negative pressure. His lungs have burst and the negative pressure helps the ventilator keep his lungs inflated.

"The baby had a drop in its hematocrit, its blood count, during the night. And there was some evidence of neurological deterioration, movements have decreased. The baby's had some seizures. This family had a previous baby with hydrocephalus. They went through two years of financial and emotional agony before the child died. They have made it very clear that they absolutely are concerned about the quality of this baby's life.

"Don, who is the attending this month, was calling to tell me this baby just had a CAT scan [a computerized three-dimensional X ray] and that it showed a very large ventricular hemorrhage. There is a chance that if we keep this baby on the respirator it will survive. But there is an extremely high likelihood of major, permanent neurologic impairment.

"So Don was asking me how I feel in this situation about turning off the respirator. He said the nurses seem comfortable with the idea. The doctors involved are comfortable. And the family wants it. If we turn it off, the baby will die. If we don't, the baby could survive with major brain damage. Now, I don't think that's playing God, because if the family said they didn't want it turned off, we wouldn't turn it off."

We spent the next hour going over Ann's and Brian's records, and then Warshaw took me down to the special-care unit and gave me a brief tour before handing me over to the nurse in charge that shift, a woman in her thirties I'll call Jean. We met Jean in the nurses' lounge, a narrow, crowded room at the end of the hall equipped with a coffee bar and a refrigerator. Several nurses were there taking a break, and Warshaw and I sipped coffee and joined in the banter and gossip. It was immediately obvious how fond they are of Warshaw. I seldom heard him addressed as "Doctor," a title many physicians insist upon with nurses and other medical personnel as a means to establish their absolute authority. Warshaw was frequently addressed as just "Joe."

In addition to running the unit and teaching, Warshaw also has research interests, and he soon left to go to another building in the medical center where he has a laboratory. Jean had finished her coffee, so she took me in tow and we entered the large room filled with isolettes and Kriselman tables where the nursing care is concentrated and the sickest babies live.

The room was about twenty by forty feet, without windows, and brightly lit by fluorescent lights. The isolettes, enclosed boxes made of transparent plastic in which the babies live, were placed in rows, interspersed with the radiant warmers. These are small tables with sides and bright lights and a heater above. The sickest infants were on these tables so they could be constantly watched and tended to. Each of these babies was attached to an array of hoses and tubes. A steady stream of nurses, interns, residents, the attending, and other physicians administered to them. A positive sign of an infant's progress is when it is moved from one of the tables to an isolette.

My first impression of the room was the sounds of technology. Almost every infant was attached to a heart monitor and each monitor beeped loudly, filling the room with high-pitched squeaks. It sounded like a pond full of soprano frogs. Some infants were also attached to apnea alarms, tiny air mattresses that trigger a steady alarm if the infant doesn't move for ten seconds. All during the day I spent in the unit, the apnea alarm on a baby girl named Jennifer repeatedly malfunctioned, filling the room with its peal. Jennifer was fine each time, but the nurse assigned to her never failed to rush to the isolette, grumbling all the while about the "damned alarm."

Jean began a methodical review of each baby in the room, stopping beside a Kriselman table or isolette to briefly describe the child's problems. When Warshaw had quickly walked me through the room earlier, I asked him which baby he had been talking about. He pointed to a corner of the room. "Over there. The table," he said. I could see two tables, so I wasn't sure which was the baby when Jean and I began our rounds.

But I knew the baby the moment we reached the table and Jean spoke.

"This little fellow's not doing so well," she said. She had lowered her voice and there was a note of sadness in it.

"Is this the one with the ventricular hemorrhage?" I said, trying to sound knowledgeable.

She shot me a quick look. "Yes. He hemorrhaged during the night and we ran a CAT scan this morning."

The premature, seriously ill infant is not a pretty creature to look at. This baby was no exception. It was extremely small, perhaps no more than twelve or fourteen inches long. It was lying on its back, with just a diaper on, its tiny legs drawn up grotesquely toward its abdomen and sticking into the air. Its skin was red.

An array of wires and tubes was attached to the baby. The lead to the cardiac monitor was taped to its chest. There was a temperature-sensing device taped to its belly, which controlled the infrared heaters above the table, keeping the baby's surroundings at a precise temperature. Thick black hoses protruded from the infant's ribs on each side. These tubes maintain the negative pressure in the chest cavity. The hoses disappeared over the side of the table to some machine below. The respirator was attached via more thick black hoses to a connector that had been inserted in the baby's nose and taped down. The respirator was on a shelf above the table. A needle behind a glass window jumped each time the machine inflated the baby's lungs. I counted about thirty of these mechanical breaths a minute.

Most of the other babies in the unit were dressed in frilly garments. Grandmothers and aunts and cousins had been busy with their knitting needles. In many of the isolettes or on the tables beside the infants there were toys—dolls, windup music boxes, squeaky rubber animals. But the Smith baby had none of these. It wore no pink or robin's-egg-blue knitted sweater. There were no toys. I noticed later that someone had given it a windup music toy, which was tucked out of the way on a shelf above the table.

Looking at this baby, I slowly began to realize I was having difficulty thinking of it as a human being. I thought of my own son, born healthy a decade before. He spent the usual three days in the nursery and then went home to begin a normal infancy. I had no trouble regarding him as a miniature person. But the form on the table didn't seem human. Was it the size and color and the spastic way it drew its legs up? Was it the technological trappings that surrounded it so completely? Or was it the fact that I knew this baby was doomed, that sometime in the coming hours the hoses would be removed and the respirator switch turned off? I pondered this question off and on all day and never satisfactorily answered it.

The rest of the babies Jean showed me were a typical mix of special-care-unit patients. One baby had hyaline membrane disease. There were two sets of twins. Just the week before, the unit had discharged a set of triplets. One twin set, Jean said, was a particularly sad case. One of the twins had received more than its share of nutrients in the womb. It would grow up normal, but its sibling would be retarded. I wondered about what this would do emotionally to the family and the normal twin over the years. I wondered if the parents would come to wish the retarded twin had been sicker and died.

There were two Rh babies and several babies that simply had been born prematurely and needed the special support of the unit. There was a baby from a Caribbean island born with a defect that pushed its intestines and stomach up into the chest. It was expected to die within a few weeks. Warshaw had told me the parents planned to take the baby home and care for it until it died. He described it as a bittersweet experience they felt they must have.

The tour completed, I took up my station near a back wall of the room, sitting on a stool, my notebook open on my knee. I wanted to watch the bustle of the room. But I soon found my gaze straying again and again to the Smith baby (its name was on a tag on the table). It wasn't long before I began prowling about the room, my path inevitably taking me past the baby. I felt self-conscious about standing and openly observing it. But some force kept drawing me toward it. As the day wore on and I became familiar with the routine, I realized that everyone in the unit was being drawn to the baby just as I was.

Several physicians were present in the unit during the day. The two interns were at the bottom of the pecking order. One, a young woman, planned to become an obstetrician-gynecologist.

She was spending several weeks in the unit before rotating on to another service in the hospital. A second intern was following the pediatrics schedule and also would rotate out in a few weeks. There was a resident, a doctor in the midst of his training to become a specialist in pediatrics. Supervising them all was the attending, a doctor who had finished his pediatrics training and had become an expert in the intricacies of neonatology. The attending, whom Warshaw had called Don, was one of two pediatricians who shared with Warshaw the responsibility of supervising the unit's day-to-day activities on a rotating basis—one month on duty and two off. The other two months were devoted to teaching, research, and care of other patients. During the day, the interns and the resident seldom left the unit. Don, the attending, was in and out.

The nurses had a busy routine. The sickest babies, such as the Smith baby, had one nurse assigned to them full-time. The nurses rotated this duty every two hours. If the nurse on duty had to leave the immediate vicinity of her infant, another nurse would stand in. Elsewhere in the room, other nurses were busy with the endless task of changing the diapers on the less-ill babies and feeding them. ("Weigh diapers" was a frequent admonishment on signs taped to the isolettes. The "input" and "output" of the infants must be carefully monitored.) Rocking chairs were standard equipment. An infant would be taken from its isolette and carried by a nurse to a rocking chair to be fed. The healthiest babies were fed from bottles. Others were fed from a device that forced food down a tube into the stomach.

There is a great emphasis in the unit upon holding the babies, cuddling them, and talking to them. That was why Rick was drawn into the circle of doctors working on Brian minutes after his delivery and told to touch his son. At one point in Brian's rocky road to good health, Ann was encouraged to hold him even though a tangle of tubes and lines was attached to him. The neonatal nurses will go to great lengths to move equipment and shift tubes and hoses so a parent can hold a baby.

As I watched the unit's routine I became aware of several parents in the unit with their children. They wore over their street clothes the same white or green smocks that the nurses and doctors wore. At first I thought they were part of the staff. One young couple, perhaps not even twenty yet, held their premature boy. The father had his long brown hair drawn into a ponytail. The mother had short blond hair. They were an island of intimate concentration as they fed their baby and then took turns simply holding him and rocking in a chair. They whispered to one another and the baby and were oblivious to their surroundings. I began to understand what Warshaw meant when he spoke of the depth of attachment parents and nurses begin to feel for the infants in the unit. The nurses cooed and clucked to the babies and offered them toys and talked baby talk to them. "Now you be good and mind your Auntie Fran," a nurse cooed to a baby named Tommy as she prepared to take a break and hand over responsibility to another nurse.

Shortly before noon, the resident spent some time with the Smith baby. From my perch on the stool I watched him as he performed a slow, methodical neurological exam. He pinched the baby's foot, tapped its knee to test the reflex point there, felt the soft area at the top of its skull (causing the baby to wriggle), and repeatedly picked up its hands, testing how well the baby could grasp his finger. He paused frequently in this procedure and stared blankly across the room, leaning his arms on the edge of the low railing that surrounds the table. He obviously was assessing the baby's condition again, making certain that there was no mistake about the baby's worsening condition, that the visible things he could measure agreed with the results of the CAT scan. Finally he reached out and pulled the two black respirator hoses loose from the connector taped to the baby's nose, in effect turning off the respirator.

The baby immediately began to squirm, its legs jerking spasmodically and its hands waving. The resident held the hoses in his hand, still leaning on the side rail, watching. Then, after perhaps fifteen seconds, during which the baby's movements grew even more agitated, he reconnected the hose. The baby's jerky movements ceased as the life-giving respiration resumed. He continued to lean on the table rail and Jean joined him. They conferred in low tones for a few minutes before he continued his rounds.

It was then I became aware that more and more of the nurses

were passing by the Smith baby's table and talking to it, leaning on the rails as the resident had and staring at it.

A young nurse, with blond hair and a rash of pimples on one cheek, spent several minutes looking at the baby. She offered her finger again and again to the little hand. She tapped the knee reflex point and pinched a foot. Then she reached up on the shelf above the table and retrieved the windup music toy, which I hadn't seen until then. It was a Mickey Mouse Club musical chime. She wound it, the key making a loud clacking sound, and placed it by the baby's head. It began playing the Mickey Mouse Club song. She walked away, leaving the toy by the infant's head, tinkling the tune until it ran down. Other nurses, some from the nursing station outside the room, some from other rooms within the unit, began to stop beside the table. They would look at the baby and touch it. Sometimes they came in pairs and studied it, talking in low whispers.

Jean paused by my stool during her rounds. She stood silently for a few seconds. We were both watching a pair of nurses standing by the Smith baby across the room. "Most of us feel okay about this," she finally said. "You see the end result in some of these cases. You know what the child would be like if it lived." She moved on.

Shortly after noon, the attending entered the room trailed by a tall, middle-aged man with a mustache and black hair graying at the temples. He wore a green pullover sweater. I knew at once that this was the baby's father. They stood for several minutes beside the table. The attending demonstrated some of the neurological tests, but he didn't remove the hoses. The father didn't touch the baby. He was very somber. They left after a few minutes, the resident steering the father by an elbow, and entered a conference room down the hall past the nurses' station.

I met Warshaw for a quick sandwich in the hospital's coffee shop. I asked him about the resident's examination of the baby, particularly about removal of the hoses.

"He was examining to what extent the baby is dependent upon the ventilator. The squirming revealed that the baby was in respiratory distress. As you can see, he needs the ventilator."

"The nurses seem to be coming by one by one to say goodbye," I ventured.

"They're doing their own tests. They want to see if they agree with the doctors. Everyone is spending the day getting used to this situation," he said.

At 2 P.M. the resident and the attending appeared with the father, who was pushing his wife in a wheelchair. She had delivered her son by caesarian section two days before and still had the pallor of a surgical patient. An intravenous-fluids bottle dangled from a pole attached to the wheelchair. Her hair was pulled back and tied with a yellow ribbon. She wore a blue bathrobe, and the hem of a yellow nightgown peeked out from under the robe. A white hospital smock had been thrown over her front. Her husband wheeled her up beside the Kriselman table and the resident lowered a rail so she could reach into the table without rising from the wheelchair. They stood around the table, staring at the baby, unspeaking, as if a painter with an eye for composition had posed them. The mother reached up and laid a finger in the baby's palm. I couldn't see whether it responded or not. "They're seeing everything in double image," Warshaw had said at lunch. "Their other baby, the one with hydrocephalus, was in here several years ago. That one they took home, but it died after two years."

They stood at the table, seemingly frozen, for four or five minutes. Then the father wheeled his wife down the hall past the nurses' station to the private baby-feeding room. The attending went in after them, carrying the baby's chart. The resident remained behind to resume his duties. While the group stood beside the baby, the others in the room had studiously avoided watching them. When they left, there was a palpable draining of tension.

At midafternoon, the nurses took up a collection and dispatched one of their number to the hospital gift shop. She returned with a sack filled with candy, cookies, potato chips, and gum. A junk food banquet followed, a ritual that occurred each afternoon. As the day progressed, the uneasiness I felt as an intruder diminished and the nurses and the physicians displayed friendliness toward me and a willingness to talk about their work. I didn't consciously attempt to steer the conversation around to the Smith baby. But the subject came up several times, not so much because they knew I was a writer and interested in it but because it was on their minds.

The Smith baby's condition had further worsened during the day, one nurse told me. "Do we give up on them or do they give up on us?" she asked. "It's funny. They always get worse when we find out."

"It's so thick in here you could cut it," the young woman who was the obstetrics-gynecology intern told me. "We've had four deaths in here in the last two days and now this."

In the late afternoon the nursing shift changed. The new nurses appeared and were briefed by those about to go home. It reminded me of the manner in which air traffic controllers hand over the airplanes they are responsible for at the moment the shift changes. I suddenly felt out of place as the familiar faces disappeared, so I moved down the hall toward Warshaw's office. I found him sitting on his couch with his feet on the coffee table, reading a medical text. After joking with him about spending all his time in such repose, I asked what would happen next with the Smith baby.

"Here again, our concern is to follow the family's wishes. One of the problems we've encountered in this is feelings of guilt on the parents' part afterward. To our surprise we discovered that the parents, even the whole family, sometimes want to be involved at the end. One of the things we offer them is a chance to be involved in the end, to hold their baby."

"Is that something this couple wants?" I asked.

"There are indications that they do," he replied.

"How will you handle the . . . uh"—I groped for a word—"the mechanics?"

"What will happen is that they probably will be in the conference room. The attending or the resident will remove the chest tubes and the respirator and the other wires and tubes and wrap the baby in a blanket and take it to them. They will be left alone with the baby as long as they want. Or if they want one of us with them, we'll do that, too."

He said nothing would happen until the evening, when the activity in the unit was settling down for the night.

I was to drive that evening from New Haven to New Canaan for dinner with Rick and Ann. We planned a discussion of their experiences with Brian in the special-care unit. I was torn between keeping that appointment and remaining in the unit. I felt certain Warshaw would let me remain, if I asked. But the idea of staying troubled me. As a journalist, I have often been offended by the insensitivity my colleagues sometimes display in situations like this one. To me, television reporters and the questions they ask the survivors at the scene of a tragedy are particularly revolting. Warshaw might permit me to remain in the unit, trusting me to remain at a discreet distance and not write something later that would violate the family's privacy, but did I really have any business watching such an intimate moment of anguish, even from across the room? I decided that I didn't.

I drove to New Canaan and was warmly received by Ann and Rick. I played with Brian until his bedtime and then sat down to a lovely meal Ann had prepared, although she apologized several times for overcooking the potatoes. Over coffee and tea we talked late into the night about Brian's two-month ordeal in the special-care unit, the two occasions when they were convinced he was about to die, and their affection for many of the doctors and nurses at Yale with whom they spent so much time. They told of their plans to donate a sizable amount of money to Yale to finance research into neonatal problems.

I drove home that night, arriving after midnight. I planned to rise early and call Warshaw before he departed at midmorning for a medical meeting in Florida. The events that I knew must be occurring in the Yale unit were on my mind as I drove.

Warshaw had said that in order to maintain a bearable emotional distance from their work, the doctors in the unit had to place qualitative values upon the young lives they dealt with. The death of a three-year-old or a twenty-year-old was qualitatively more of a loss than the death of an 800-gram baby born twelve weeks too early.

Does that mean that as you move down the scale toward the moment of fertilization, life has less and less value? Does a thirty-two-week Rh fetus have more value than an eighteen-week Down's-syndrome fetus whose mother has decided upon an abortion? Is the blastocyst formed within hours of the union of egg and sperm more valuable than the egg and sperm before the moment of fertilization? Was Baby Boy Smith's life of less value because he would be severely retarded than that of a similar baby who would grow up with a normal brain? The Smiths and their doctors could decide to turn off the baby's machines. But suppose they didn't and he lived to, say, age five when he needed a life-saving operation. Could they then refuse the surgery, condemning their child to death? I doubted society would permit that.

Such questions have been long debated by philosophers and theologians and doubtless will be debated for years to come. But they reflect real problems plaguing doctors and their patients. The opposing forces in the abortion issue are drawn up on both sides of a chasm that sometimes seems impossible to bridge. The question of when to turn off the life-support machines is equally pressing, and the answers are even less founded in law.

I had wondered several times how doctors such as Berkowitz and Hobbins could struggle against enormous odds to bring an infant like Brian into the world and at the same time care for other patients whose needs dictate abortion. How do they reconcile this giving and taking of life—if indeed abortion is a taking of life, a contention I basically disagree with.

The answer that began to emerge was that doctors aren't playing God as is often claimed, but instead are merely technicians, mechanics, ready to do their clients' bidding. Perhaps a generation ago they played God, but modern doctors have changed. They are growing more sensitive to their patients' needs. Joe Warshaw and his colleagues might have kept the Smith baby alive. Had the infant survived, its parents eventually could have taken it home, if that had been their wish. Few people would argue with that notion. But if the Smith family has that freedom, then does a woman have the right to seek an abortion simply because she has no desire to have a baby? Does the couple determined to produce a girl have the right to undergo amniocentesis with each pregnancy and abort all male fetuses?

The more we learn to control the environment through science and technology, the more we can alter people's destinies. This ability raises a plethora of moral issues that no one has dealt with definitively.

When I telephoned the next morning, Warshaw picked up his phone on the first ring.

"I haven't talked to Don about it yet, but I think it went pretty much as I predicted," he told me. He sounded brisk.

I was brimming with a reporter's questions. What time did the baby die? Who unhooked the hoses and carried the baby to the parents? Did the baby thrash about when the hoses were removed? How long were they in the room with the baby? Was a doctor with them? Did the baby die in its mother's arms?

But I didn't ask. I sensed that Warshaw really didn't want to talk about it. Part of the reason was that he was hurrying to clear his desk and get out of the office. But part might be that unconsciously he felt I had seen enough, intruded enough. Some things between the physician and the patient are sacrosanct, even if the patient's identity is disguised.

Some things are too painful for anyone to talk about. Ann and Rick taught me that the night before. At one point during dinner I told them about what I had witnessed in the same room where they virtually lived for two months with Brian. There was a shocked silence. Someone quickly changed the subject.

I realized there are no answers. Each person makes his own peace with these issues in the depths of his soul. Each case must be judged on its merits.

But who is to be the judge?

AN AFTERNOON IN SEATTLE

One winter afternoon in Seattle, as low, gray clouds glowered down at the city, I sat on a metal folding chair in the sparsely furnished office occupied by a woman who is a "clinic coordinator" in the University of Washington's genetics program. I'll call her Janet. I had come to talk about Charlie and Arlene Scott and their adopted daughter, Christine. But our conversation soon veered off into talk of how families deal emotionally with their problems. Janet spoke of the guilt, transferral of blame, denial, and all the other forms of coping that I had seen already and would later see in still more families.

I told her a story Joe Warshaw at Yale had recounted. A deformed baby was born to a Latin-American immigrant couple living in a central Connecticut industrial community. The seriously ill infant was brought to Yale's neonatal special-care unit and the couple ultimately referred for genetics counseling. The father told the genetics counselors that the family's physician had explained the cause of their problem: weak sperm!

Janet then recounted the story of a woman she had worked with who had given birth to a Down's-syndrome child ten years earlier. The infant had been institutionalized at an early age. A decade later, the mother asked that her normal teenage daughter receive genetics counseling, just in case the daughter might be at risk for producing retarded children, too. The geneticists suggested the institutionalized child also receive chromosome studies. When the woman appeared and met with the doctors to

learn the results of the studies, she confessed that for ten years
she had blamed herself for the Down's child's misfortune. While
pregnant, the woman had gone horseback riding and fallen off.
For a decade she believed that the fall had shaken her chromosomes loose! In fact, the woman told the Seattle geneticists, a
doctor had told her that.

The Hispanic man did not have weak sperm, of course. It is
conceivable that a physician whose training never included medical genetics might tell a patient he had weak sperm, a particularly damning indictment for a man whose cultural heritage is so
entwined with feelings of *machismo*. It is equally conceivable
that a doctor, doing his best to explain about genes and eggs and
sperm and dominant and recessive traits, might lead a patient to
conclude that he had weak sperm.

The fall from the horse didn't knock the woman's chromosomes loose, either. Perhaps a doctor did tell her such a thing.
But when the Seattle doctors told her that couldn't possibly have
happened, she had a long-festering emotional burden lifted.

It's clear that to far too many minds genetics presents a mystery so deep that the notion of retarded sperm and loose chromosomes seems as plausible as any. This aura of mystery adds to the
difficulty of communications between the genetics professionals
and the public. It's not simply a matter of teaching a patient
with a hereditary defect the basics of genetics, as if the subject
were how to reupholster furniture or prepare an income tax return. People who need such genetics instruction are plunging at
the same time through a storm of emotion. Guilt, fear, recrimi-
nation, resentment, and even hatred are swirling over the patient
sitting in the doctor's office while the awful truth is pronounced.
As the geneticist earnestly outlines the basics of the specialty,
the patient's mind is in another realm, trying to deal with very
fundamental questions of self-image. It is as if a part of him or
her has died and all the rituals of mourning that we know are so
necessary to the psyche must be performed. Only then—days,
weeks, months, or even years later—is someone ready for instruction. Only then is there a chance that he or she can hear and understand the facts well enough to make informed decisions.

In all the families whose destinies have been altered by their
genes and who poured out their stories to me, I see a common

difficulty: coping emotionally with their problem and truly grasping its meaning. Clearly, it is not enough for the geneticist to take medical histories, construct pedigrees, perform diagnostic tests, assess all the data, and then deliver the genetic facts based on the best knowledge available. Patients need more. They need time. They need to have the problem explained over and over again. They need frequent reassurance. Most of all, they need to talk—to find a knowledgeable, understanding, neutral party. Every genetics patient should have, like Marsha, the mother of the Tay-Sachs baby, a Sandy Silverman, standing always ready to give emotional support as long and as often as necessary.

"I think when you learn that you have a genetic problem, it's something that's with you forever afterward. It will always be there in your subconscious," Janet said to me that afternoon in Seattle. I nodded my head, encouraging her to continue.

"You see," she blurted, "I have a genetic disease myself."

I listened with rapt attention as she told her story. A couple of years before, her mother and sister had come to Seattle for a visit. As a clinic coordinator, Janet was learning her genetics on the job. Her mother had always had a problem with the muscles in her legs, in climbing stairs. Janet began to wonder if her mother's difficulty and her sister's similar problem could be genetic. She arranged for blood tests. The results confirmed that her mother and sister have a form of muscular dystrophy that affects the leg muscles.

"I said to myself, 'Gee, I wonder if I'm minimally affected?' I was starting to have problems climbing stairs, too. I would get tired. But you know. You're out of shape. You're ten pounds overweight, so you don't think anything about it. So finally I worked up my nerve enough to take a blood sample down to the lab for a CPK analysis. Then I forgot about it."

But several weeks later, when she remembered and called the lab, Janet was stunned to learn she had the same disease her mother had. She was minimally affected, and the onset was at least ten years later in life than her mother's and sister's disease. But with certainty, muscle problems in her legs would worsen. She was devastated.

"It's something that never really leaves your mind. It's something you try not to dwell on, but you look at everything differently. When you go for a hike, you look at it differently. It's a bittersweet thing because you know you're not going to be able to do that much longer. Water skiing I miss very much. I have water-skied every summer for ten years. Now no more. A genetic disease affects your outlook on life, how you're going to live your life, the values you place on things, whether to have children.

"In terms of my work, it's been an interesting experience. In the last year there have been three families that I've met with to talk about their experiences before or after being in the genetics clinic. They've said to me, 'Well, you wouldn't understand because you don't have a genetic disease.' It's really been interesting to tell them, 'Oh yes I do. I know what you're talking about.'

"All of a sudden, a whole barrier falls away. They relate to me in a totally different way. They're just amazed that somebody understands how they feel."

Looking back on that conversation, now I see that Janet had found the answer to communicating with the victims of hereditary conditions: understanding how they feel. That is the key—for doctors, counselors, educators, and all who are convinced that the public must move quickly to understand and accept the coming explosion of genetics knowledge that will thrust ever more complicated moral and social dilemmas upon us.

When I look back upon "my" families with genetic problems, I see that those who have dealt with their disease most handily also understand the situation more thoroughly. They are the families who developed relationships with a doctor or a genetics counselor and received as much attention as they needed.

Marsha and Gary had Sandy Silverman. Sandy probably played a bigger role in holding Marsha and Gary together and helping them weather their crisis than she realized.

Judy Derstine, the genetics counselor in San Francisco, was particularly close to Sandy all through her ordeal.

Ann and Rick were unusually fortunate in that their doctor became a close friend and was, in addition, warm and approachable.

But the families who seem to have coped not well at all lack that neutral, always available ear. The Buckinghams are particularly isolated. The O'Caseys either have no one to turn to or are unwilling to do so. Their fundamental lack of understanding about HLA antigens (always a complex subject to begin with) and how they have touched their lives is a major factor in their situation.

Perhaps Victor McKusick, the famed Johns Hopkins medical geneticist, was not after all brushing aside the questioning doctor on the afternoon I attended the genetics short-course clinic in Bar Harbor, Maine. ". . . You work the parents in on the problem slowly," McKusick had answered when the questioner kept pressing. "You don't hit them all at once. You bring along their education in this matter."

What I had heard then—and it took me several months to realize it—was a medical geneticist speaking in professional shorthand. You work with the victim of hereditary disease step by step, McKusick was saying. Implicit in that approach is committing the time to meet the patient's needs fully. It is clear to me now that we need more Sandy Silvermans, Judy Derstines, and Janets helping those with genetically altered destinies deal with life.

Janet should have the last word:

"I know from talking to people who have been through our genetics clinic that just because they can look at you without crying and ask the right kinds of questions doesn't mean that it's any less devastating to them when the door shuts behind them and they're home."